Francis Fontaine

The State of Georgia

What it offers to immigrants, capitalists, producers and manufacturers, fruit and

vegetable growers, and those desiring to better their condition. Vol. 1

Francis Fontaine

The State of Georgia
What it offers to immigrants, capitalists, producers and manufacturers, fruit and vegetable growers, and those desiring to better their condition. Vol. 1

ISBN/EAN: 9783337373665

Printed in Europe, USA, Canada, Australia, Japan

Cover: Foto ©berggeist007 / pixelio.de

More available books at **www.hansebooks.com**

THE
STATE OF GEORGIA:

WHAT IT OFFERS TO

IMMIGRANTS, CAPITALISTS, PRODUCERS AND MANUFACTURERS, FRUIT AND VEGETABLE GROWERS, AND THOSE DESIRING TO BETTER THEIR CONDITION.

BY

FRANCIS FONTAINE,

STATE COMMISSIONER OF LAND AND IMMIGRATION.

WITH ILLUSTRATIONS.

ATLANTA, GEORGIA.

1881.

PREFACE.

The following letters are offered the Public as a suitable preface to this pamphlet, in which testimony from Northern writers is copiously presented that truth without exaggeration may be given.

Correspondence Augusta, (Ga.,) Chronicle.

NEW YORK, May 5th, 1880.

"The comparative growth of the thirteen original States, presents a record that is hard to realize. But, as evidence of the strange truthfulness of figures, compare New York and Georgia, the two that exhibit the highest rate per cent. of increase, and this is the result: In the 80 years from 1790 to 1870, New York increased in population 12 fold, or 11.88 per cent., while Georgia increased 14 fold, or 13.44 per cent., the highest of any of the Atlantic States. That, too, with the greatest drawback as to distance from centres of population and trade, and in being out of the line of immigration. But, as a still higher proof of the vitality of the State, take the returns for 1870, including the five years of the war. During that time Maine and New Hampshire both showed a decrease in population; but, in spite of a bloody and desolating war, partly within her own borders, in spite of the loss of life and property, in spite of what has been called "the indirect loss by the war in the check given to the increase of the native population," and the total cessation of immigration, Georgia showed an increase of 8 per cent. in her white population. The losses by death and disease in the Union army were put at 500,000, while those of the Confederacy amounted to 350,000, of which Georgia bore her share. These facts, taken in connection with the depressed condition of affairs immediately succeeding the war, make any increase of the whites most gratifying, and to rebuke the cry of persecution to the negro the record is that while there was only an increase of 9 per cent. throughout the entire Union, in Georgia there was a growth of 17 per cent. of this colored ' bone and muscle ' of the country."

General Wm. T. Sherman, of the United States Army, thus wrote to the Editor of the Atlanta, (Ga.,) *Constitution*, in 1879:

" Now that slavery is removed, there is no longer any reason why Georgia, especially the Northern part, should not rapidly regain her prominence among the great States of our Union. I know that no section is more favored in climate, health, soil, minerals, water, and everything which man needs for his material wants, and to contribute to his physical and intellectual development. Your railroads, already finished, giving your people cheap supplies, and the means of sending in every section their surplus products of the soil or of manufactures. You have immense beds of iron and coal, besides inexhaustible quantities of timber, oak, hickory, beech, poplar, pine, etc., so necessary in modern factories, and which are becoming scarce in other sections of our busy country.

" North Georgia is peculiarly adapted to fruit orchards, to gardens and small farms, and all you need to make it teem with prosperity, is more people from that class of Northern farmers and manufacturers, and that other large class of European emigrants which has converted the great Northwest from a wilderness into comfortable homes for its millions of contented people.

"I have crossed this continent many times, by almost every possible route, and feel certain that at this time no single region holds out as strong inducements for industrious emigrants, as that from Lynchburg, Virginia, to Huntsville, Alabama, right and left, embracing the mountain ranges and intervening valleys, especially East Tennessee, North Georgia and Alabama. I hope I will not give offense in saying that the present population has not done full justice to this naturally beautiful and most favored region of our country, and that two or three millions of people could be diverted from the great West to this region with profit and advantage to all concerned. This whole region, though called 'Southern,' is, in fact, 'Northern,' viz.: it is a wheat-growing country; has a climate in no sense tropical or Southern, but was designed by nature for small farms and not for large plantations. In the region I have named, North Georgia forms a most important part, and your city, Atlanta, is its natural center or capital. It is admirably situated, a thousand feet above the sea, healthy, with abundance of the purest water, and with granite, limestone, sandstone and clay convenient to build a second London. In 1864, my army, composed of near a hundred thousand men, all accustomed to a Northern climate, were grouped about Atlanta from June to November without tents, and were as vigorous, healthy and strong, as though they were in Ohio or New York. Indeed, the whole country from the Tennessee to the Ocmulgee, is famous for health, pure water, abundant timber, and with a large proportion of good soil, especially in the valleys, and all you need is more people of the right sort.

"I am satisfied from my recent visit, that Northern professional men, manufacturers, mechanics and farmers, may come to Atlanta, Rome and Chattanooga with a certainty of fair dealing and fair encouragement. Though I was personally regarded the *bete-noir* of the late war in your region, the author of all your woes, yet I admit that I have just passed over the very ground desolated by the civil war, and have received everywhere nothing but kind and courteous treatment from the highest to the lowest, and I heard of no violence to others for opinions' sake. Some Union men spoke to me of social ostracism, but I saw nothing of it, and even if it does exist, it must disappear with the present generation.

"Therefore, I shall believe and maintain that North Georgia is now in a condition to invite emigration from the Northern States of our Union and from Europe, and all parties concerned should advertise widely the great inducements your region holds out to the industrious and frugal of all lands."

The following letter describes with equal truth the coast counties of Southeastern Georgia:

CAMDEN COUNTY.

"I am a native of New York State; did mercantile business for twenty years at Waverly, Tioga county, in that State; came to Georgia for my health in 1869. Since my residence here, nearly ten years, myself and family have enjoyed uninterrupted health, winter and summer. From my experience and observation, I believe the climate of the Southern coast of Georgia cannot be surpassed for health and comfort during the entire year.

"The soil, with proper culture, will produce every variety of vegetables, and is most grateful for kind treatment. Even with inferior cultivation, the soil yields a return that could not be realized in the most favored locality in the North, under the same treatment. For growing the orange, or any other semi-tropical fruit grown in Florida (north of the frost line,) the Southern coast of Georgia, for sixty miles, has advantages over the orange district 100 miles South. The orange *tree* is more hardy, less liable to injury from cold, and the fruit has a thinner skin and higher flavor. I have 1,500 trees. Not a single year old seedling killed by the cold last January, while the trees in central and middle Florida suffered serious injury. Farm crops successfully grown are cotton, corn, sugar cane, sorghum, peas and beans, Irish and sweet potatoes, oats, rye, etc.

"This region is far more healthy than any section of the North or West with which I am acquainted, and we have at St. Mary's as peaceable and law-abiding class of people, white and black, as can be found in any section. So far as I have seen, there is less sectional feeling in the South than in the North, and I have never had any fear of personal violence to myself, family or to any Northern man who may desire to settle in Georgia. For nearly ten years that I have lived South, I have, without exception, received the kindest treatment and evidences of good will.

"The men who now represent the condition of society at the South to be such as should deter a Northern man from settling here, are enemies to the poor, white and black, North and South. Such men, who still appeal to the passions, were not clothed in blue or gray (during civil strife) for honest purposes; if wearing either color they were the *home guards*, or men seeking some personal benefit or political position. I have no doubt the persistent misrepresentations of the Southern people, has deterred many good men from seeking homes in the South; who, could they have known the truth, would now be in the possession and enjoyment of free and independent homes in the South, freed from the anxieties of their present condition North.

"Taxes in Pennsylvania and New York, where I have real estate interests, are as four to one in Georgia. In Georgia, taxes are low on a very low valuation; in the North, they are high on a high valuation.

"If all Georgians would work for Georgia as the Floridians work for Florida, the population would be doubled in ten years.

"In my opinion there is no State in the Union that has the undeveloped wealth of Georgia. Every variety of fruit and grain grown in the United States can be successfully grown in Georgia; its mineral wealth is very great, and its advantages for manufacturing everything useful are unsurpassed. Every variety of climate, from the balmy air on its Southern coast, to its mountains and snow of winter in the Northern portion.

"I am proud of my native State, New York, but equally as much interested in the prosperity and full development of my adopted State South. * * * *

"SILAS FORDHAM,
"St. Mary's, Camden County, Ga."

Dr. George Little, the State Geologist, in his report for 1875, says:
"Every fruit and cereal, and textile fibre useful to man, can be cultivated in one portion or another of the State. Every variety of climate is afforded, as illustrated in my own experience during the present month, when leaving one party on the Southern border sleeping in the open air on the islands of Okefinokee, with oranges and bananas hanging in the gardens on its borders. I joined in the same week another party on the Cohutta mountains covered with snow; while in passing through Atlanta, balmy breezes were blowing as if it were spring-time."

The object of this publication is to give official and reliable information in concise form to those who contemplate seeking new homes upon cheap lands of virgin fertility. Georgia, though a Southern State, is not tropical, but the entire State lies in the temperate zone, and is, by common consent, recognized as the "Empire State of the South." I have purposely copied *verbatim*, whenever accuracy was not sacrificed to brevity, the reports made to the Georgia Department of Agriculture, in order that every statement made shall be official, and that facts rather than theories, shall especially characterize the work. For some of the illustrations and some important suggestions, I am indebted to the reliable "Guide to Georgia," by Prof. J. T. Derry, of Georgia. My acknowledgements are also due Mr. J. T. Henderson, Commissioner of Agriculture of Georgia, the Savannah, Florida and Western, and the Macon and Brunswick Railroads, and Messrs. W. T. McArthur and Wm. E. Dodge of the Georgia Land and Lumber Company, for courtesies and information extended.

FRANCIS FONTAINE,
Commissioner of Land and Immigration.

ATLANTA, Ga., July 29, 1880.

CLIMATE AND TEMPERATURE.

Statistics show that the population of Georgia differed from that of Michigan in 1870, only fifty souls, and the vital statistics show that their death rate is about the same.

The yearly death rate in Georgia,   is 1 to 88 inhabitants.
      "   "   Illinois,   " 1 " 73   "
      "   "   Connecticut,   " 1 " 74   "
      "   "   Maine,   " 1 " 77   "
      "   "   Missouri,   " 1 " 51   "
      "   "   Sweden,   " 1 " 50   "
      "   "   Denmark,   " 1 " 46   "
      "   "   Great Britain, " 1 " 46   "

"The mean annual temperature of Atlanta, is the same as that of Washington City, St. Louis, Missouri, and Louisville, Kentucky. The mean annual temperature in Southern Georgia, is 64° to 68°; in upper Georgia it is between 52° and 56°; while in the mountains it is 52°. The mean of Hall and Habersham counties corresponds with that of central Ohio, Indiana, Illinois, upper Missouri and lower Nebraska."

The following is a synopsis of the weather at Atlanta, Georgia, for four winters, with rainfall:

TABLE.

MONTH AND YEAR.	Temperature.			Rainfall	Rainy Days
	Maximum	Minimum	Mean		
December, 1876	62	11	37.5	4.10	9
January, 1877	68	6	45.5	5.93	15
February, 1877	68	25	49.2	3.10	5
For the Winter	--	--	44.0	13.13	29
December, 1877	70	20	48.0	4.11	8
January, 1878	63	17	40.3	6.11	6
February, 1878	68	22	43.6	3.30	9
For the Winter	--	--	44.0	13.52	23
December, 1878	64	17	38.4	4.87	11
January, 1879	75	7	42.2	3.84	9
February, 1879	69	21	42.0	2.72	7
For the Winter	--	--	40.9	11.43	27
December, 1879	70	16	49.6	7.20	11
January, 1880	71	30	52.8	4.46	12
February, 1880	74	28	46.8	5.11	7
For the Winter	--	--	49.7	16.77	30

At Columbus, Ga., the thermometer for the past seven years has averaged each cotton season of twelve months sixty-five to sixty-eight degrees, and the rain fall from over fifty-one to fifty-eight inches.

Snow falls in Northeast Georgia where the average elevation is 1,500 feet, and the mountains attain a height of 5,000 feet, usually from two to three times a year, to a depth varying from two inches to six inches. In Middle Georgia snow falls about once in three years, the depth varying from one-half to four inches. In Southern Georgia, snow is rarely seen, and never sufficient to remain on the ground a day. The mercury seldom rises above 90°, or falls below 32°. The following is taken from Derry's Guide to Georgia.

"From the Meteorogical Register kept by the officers of the Medical Staff at the United States Arsenal in Summerville, near Augusta, made at Sunrise, at nine o'clock A. M., three o'clock P. M., and nine o'clock P. M., for more than twenty years, we learn that the mean average temperature of the year at that point is 64° Fahr., and the mean monthly temperature to be as follows:

"For January,	46°, 7' Fahr.	For February,	50°, 7'	Fahr.
" March,	58°, 8' "	" April,	65°, 1'	"
" May,	72°, 2' "	" June,	79°,	"
" July,	80°, 9' "	" August,	79°, 7'	"
" Sept.	72°, 8' "	" October,	63°, 5'	"
" Nov.	53°, 8' "	" December	46°, 3'	"

"The mean temperature for the four seasons is shown to be, for the Spring, 65°, 3'; for the Summer, 79°, 9'; Autumn, 63°, 4'; Winter, 47°, 9'. The rain fall for the four seasons is, for the Spring, 10,16 inches; Summer, 14.14 inches; Autumn, 6.95 inches; Winter, 5.92 inches. The mean number of fair days during the year is 238; cloudy days, 127; rainy days, 70; snow, about two days in three years."

The altitude of Thomasville, (Southern Georgia), is 330 feet above the sea, and the mean animal temperature in Winter is about 53°, and in Summer about 83°, with the barometer at about 20½°. The healing influence of the pine forests blesses Middle and Southern Georgia where there are no extreme vicissitudes of temperature. Dr. T. S. Hopkins, writes: "I have before me the report of the thermometer for Thomasville, Georgia, and Santa Barbara, California, for the month of January, 1875, as follows: At Thomasville the monthly mean temperature was 55°, 50'; highest temperature, 72°; lowest temperature, 38°. Santa Barbara, monthly mean temperature, 53°, 50'; highest temperature, 70°; lowest temperature 38°. In temperature, you will perceive, we have the advantage of Santa Barbara, while in the number of fair days we know of no region that can report more favorably."

Eastman, Dodge Co., Georgia, is quite equal to Thomasville, Georgia, or Aiken, S. C., as a winter sanitarium for people suffering from catarrh, bronchial and pulmonary diseases.

St. Paul, Minnesota, has a mean annual temperature of 44°, 6'. The annual mean temperature of Winnepeg is 34°, 38', being over ten degrees colder on an average, taking the year round, than St. Paul, and eight degrees colder than Montreal.   But beside this normal and necessary physical inferiority, resulting from a colder climate, a Mennonite farmer in Minnesota writes, December 5, 1879, as follows :

"All wood required for building houses or fences is brought to my place of residence by railway, a distance of 135 to 200 miles.   Hay is used as fuel.   The Winter is long and severe, and cattle must be fed for at least six months of the year."

The cheapest and best timber in the United States is in the State of Georgia, and no farm laborer has ever been charged for fuel. About 60 per cent. of the original forest growth is still standing, or 22,200,000 acres.

The following comparisons are here offered to show the climatic advantages enjoyed by Georgia over the Northwestern States, which receive, it is alleged, over 80 per cent. of the immense immigration now coming to America.

Mr. Henry G. Vennor, of Montreal, writes to the *Albany Argus* as follows :

"We are again in Midwinter in Montreal, (8th March), with the thermometer three degrees below zero, a cutting wind and the best sleighing of the winter.   The 5th, 6th and 7th days' prediction has been certified to the letter.   The 5th gave a furious snow storm, and Quebec is yet blockaded."

The Valdosta (Southern Georgia,) *Times*, of March 6th, thus describes the situation there :

"Strawberries in the middle of January; oats heading the first of March; sugar cane growing the entire winter and tasseling in March; pepper plant exposed, green through the entire winter, and green pods first of March; corn tasseling and in silk in February; English peas and Irish potatoes from the garden in February; thermometer 83° first week in March."

We may add that the thermometer in the middle of Summer will show a more pleasant temperature than in regions further north, and that malarial diseases are not prevalent except on some of the water courses.

TEMPERATURE TABLES.

The following tables indicate the temperature at the places and for the times named :—

MONTHLY RAINFALL AT WEST END, NEAR ATLANTA.

Showing the number of days on which rain fell in each month, and the quantity that fell (in inches and decimals) in each month, from July 1870, to October 1876, inclusive, taken by Major S. B. Wight. Lat. 33° 54' North; Long. 7° 28' West from Washington. Altitude, 1084 feet above the level of the sea.

	1870.		1871.		1872.		1873.		1874.		1875.		1876.	
	No. of Days in which rain fell.	Amount of Rain-fall.	No. of Days in which rain fell.	Amount of Rain-fall.	No. of Days in which rain fell.	Amount of Rain-fall.	No. of Days in which rain fell.	Amount of Rain-fall.	No. of Days in which rain fell.	Amount of Rain-fall.	No. of Days in which rain fell.	Amount of Rain-fall.	No. of Days in which rain fell.	Amount of Rain-fall.
January	..	..	4	2.08	4	2.94	6	3.36	4	3.14	11	5.60	6	3.32
February	..	..	6	6.20	9	5.28	6	12.04	5	6.86	7	6.92	9	5.37
March	..	..	7	6.11	6	7.66	5	2.58	10	7.38	11	10.27	6	5.59
April	..	..	7	5.20	7	3.09	4	1.96	12	10.42	7	4.79	7	6.01
May	..	..	10	7.77	7	3.75	9	6.05	2	3.00	5	1.84	10	5.00
June	..	..	13	5.97	5	1.82	9	6.86	13	7.71	8	4.58	..	3.25
July	10	2.25	5	1.12	14	3.91	9	3.87	9	4.70	8	3.84	9	3.49
August	12	4.69	5	6.49	5	5.84	5	2.08	9	10.00	7	3.42	9	5.32
September	5	0.40	4	4.44	4	2.26	4	5.40	5	0.47	6	4.64	4	0.68
October	4	0.67	6	2.09	4	0.74	2	1.23	3	0.80	5	1.50	..	1.81
November	8	5.42	8	3.41	5	2.12	6	3.15	9	3.19	7	3.45	..	..
December	5	3.74	9	3.36	5	4.48	5	2.41	11	3.00	11	6.14	..	..
Totals	44	26.17	84	54.09	75	43.89	70	50.99	92	60.67	93	56.99	..	..

Average for 5 years : rained 83 days per year, and 58.33 inches fell per year.

MONTHLY RAINFALL AT AUGUSTA, SAVANNAH, AND TYBEE ISLAND LIGHTHOUSE, FROM MAY 1874, TO JUNE 1875, INCLUSIVE, AS REPORTED BY THE U. S. SIGNAL SERVICE BUREAU.

Months.	Augusta.	Savannah.	Tybee.
May, 1874	3.83	4.85	....
June, 1874	3.29	4.85	....
July, 1874	5.35	10.14	4.55
August, 1874	6.81	6.58	3.68
September, 1874	5.85	8.89	5.90
October, 1874	1.09	1.42	1.23
November, 1874	2.21	1.80	1.65
December, 1874	4.04	1.60	1.41
January, 1875	6.77	8.84	6.02
February, 1875	5.11	3.50	3.16
March, 1875	11.89	6.98	6.96
April, 1875	4.71	5.11	3.54
May, 1875	1.10	3.20	1.43
June, 1875	6.59	4.10	3.12
Totals for the time	61.51	63.12	41.95

MONTHLY RAINFALL AT COLUMBUS, GA., EXPRESSED IN INCHES AND DECIMALS, FROM JUNE 1874, TO OCTOBER 1876, INCLUSIVE. TAKEN BY DR. E. C. HOOD.

	1874.	1875.	1876.
January	....	5.05	2.79
February	....	5.57	4.16
March	....	12.34	7.90
April	....	7.57	9.17
May	....	2.80	4.45
June	9.72	2.07	4.81
July	10.50	2.25	3.50
August	1.41	6.41	5.31
September	2.29	3.09	0.62
October	0.19	5.99	3.96
November	2.69	4.66	....
December	6.51	3.88	....
Total	....	61.68	....

Greatest quantity in any week, 4.88.

TABLE SHOWING THE MONTHLY MAXIMUM AND MINIMUM TEMPERATURES, ALSO THE MEAN MAXIMUM AND MEAN MINIMUM, AND GENERAL MEAN, AT MACON GA., FROM JANUARY 1871, TO OCTOBER 1876, INCLUSIVE, AS RECORDED BY MR. J. M. BOARDMAN, MACON, GA.

Month.	1871. Maximum	Minimum	Mean Max'm.	Mean Min'm.	General Mean.	1872. Maximum	Minimum	Mean Max'm.	Mean Min'm.	General Mean.	1873. Maximum	Minimum	Mean Max'm.	Mean Min'm.	General Mean.	1874. Maximum	Minimum	Mean Max'm.	Mean Min'm.	General Mean.	1875. Maximum	Minimum	Mean Max'm.	Mean Min'm.	General Mean.	1876. Maximum	Minimum	Mean Max'm.	Mean Min'n.	General Mean.
January	70°	31°	56°	36°	46°	68°	22°	49°	29°	39°	66°	23°	50°	36°	43°	72°	28°	57°	43°	50°	70°	20°	52°	39°	46°	76°	31°	62°	43°	53°
February	75°	33°	60°	48°	54°	70°	32°	57°	40°	48°	74°	30°	69°	43°	51°	76°	30°	57°	43°	50°	78°	18°	56°	36°	46°	75°	23°	60°	42°	51°
March	81°	39°	70°	55°	63°	78°	34°	60°	44°	52°	77°	33°	61°	41°	51°	73°	38°	66°	48°	57°	79°	30°	65°	46°	55°	77°	24°	65°	43°	54°
April	84°	50°	74°	62°	68°	90°	45°	74°	60°	67°	86°	40°	73°	51°	62°	83°	40°	71°	52°	62°	80°	36°	72°	49°	60°	88°	40°	74°	53°	64°
May	91°	51°	80°	64°	72°	90°	56°	85°	70°	77°	90°	40°	80°	62°	71°	92°	49°	80°	59°	69°	89°	48°	80°	60°	70°	92°	46°	81°	65°	73°
June	91°	51°	83°	70°	78°	93°	71°	85°	72°	79°	90°	64°	83°	71°	77°	96°	70°	87°	67°	77°	98°	63°	87°	72°	79°	95°	64°	85°	72°	78°
July	94°	70°	85°	75°	80°	96°	72°	91°	73°	82°	94°	73°	86°	72°	79°	94°	70°	86°	69°	78°	98°	74°	93°	77°	85°	95°	61°	90°	76°	83°
August	95°	72°	86°	76°	81°	96°	70°	86°	75°	81°	92°	70°	85°	71°	78°	97°	70°	87°	70°	78°	98°	66°	89°	72°	77°	97°	70°	87°	75°	82°
September	85°	49°	83°	57°	70°	92°	65°	82°	69°	76°	92°	64°	81°	66°	73°	87°	62°	81°	65°	73°	92°	57°	82°	68°	74°	92°	60°	83°	67°	75°
October	83°	46°	72°	64°	68°	83°	44°	71°	53°	62°	86°	31°	71°	50°	60°	81°	40°	72°	50°	61°	77°	33°	68°	58°	63°	78°	34°	83°	66°	75°
November	76°	40°	62°	48°	55°	72°	30°	58°	42°	50°	78°	24°	60°	42°	51°	79°	26°	66°	46°	56°	78°	33°	64°	58°	57°	...	...			
December	66°	21°	51°	38°	45°	66°	30°	50°	35°	43°	72°	22°	55°	41°	48°	74°	26°	58°	40°	40°	77°	18°	60°	45°	53°	...	...			
Means for the Year	82°	46°	72°	58°	65°	83°	47°	69°	55°	63°	83°	43°	70°	54	62°	84°	45°	72°	54°	63°	84°	41°	71°	56°						
Gen. Mean for the Year					65°					63°					62°					63°					64°					

RECUPERATIVE POWER AND FINANCIAL CONDITION.

THE NEGRO IN GEORGIA IN 1865 AND 1870. ENERGY OF THE PEOPLE.

The recuperation from the losses caused by the late war between the States, has been as marked and rapid in the South as in France, since the Franco-German war. Germany exacted from France a war tax of five milliards of francs, but the cost, in francs, of emancipation alone to the Southern States was ten milliards. To this must be added the * "the tax on cotton" levied in 1866-67 by the General Government amounting to $60,000,000. Yet the finances of Georgia to-day show as good condition as that of any State in the American Union, and Georgia is the only State which "floats" a four per cent. bond which commands in the markets more than par value. The population of the ten cotton-growing States is less than twenty percentum of the population of the United States, but the balance of trade last year in favor of the ten Southern States amounted so $100,000,000. For their cotton alone, the ten Southern States are credited with $162,304,250 besides having supplied the United States, while the total value of the domestic exports of the United States for the year ending June 30th, 1879., was but $717,093,777.

Mr. John Bright thus wrote to a citizen of Georgia recently:

"As for the South, you will have few Englishmen settle there so long as the old temper of your people continues to exist. We hear of the ill-treatment of the negro, and of the hostile disposition of your white people toward families who come from the North."

The fact is, as hundreds of Northern settlers in every part of Georgia attest, Northern immigrants are cordially welcomed; and as to the negro, let the following statement demonstrate that the great Liberal leader is in error. The negro freedmen of Georgia, who started in life as penniless laborers fifteen years ago, without capital and without experience, *now own enough land to give six and one-tenth acres to each voter in the State.* In the United States Senate, the Committee on the depression of labor, reported as follows:

"Particular examination into the condition of colored men in the South, as disclosed by the testimony of both whites and blacks, republicans and democrats, showed the average compensation to be quite equal if not better than the average in any Northern State."

From the report of the Comptroller General, we copy the following exact figures:

---

* This tax has been decided unconstitutional by the Supreme Court of the United States.

No. of negro polls in the State	88,522
No. of acres of land owned by negroes	541,199
City or town property, value	$1,094,435
Money and solvent debts	73,253
House and kitchen furniture	448,713
Value of hogs, horses, cows, etc	1,704,230
Farming and mechanical tools	143,258
Other property not including crops, provisions, etc	369,751
Value of whole property	5,182,398

These returns were made on the oaths of the negroes themselves. We may be sure that there is no exaggeration, for neither negroes nor others are likely to put too high a value on property which is to be taxed according to valuation. The increase in the number of acres returned in 1879, over the return of 1878, is thirty-nine thousand three hundred and nine.

According to the census of 1870, there were in Georgia 88,522 negro polls, while the taxable value (which is less than the real value) of the property owned by them was $5,182,398. In 1865 they were penniless. According to the same census there are 888,081 producers in Georgia, and the value of the cotton crop produced by them last year is more than $30,000,000.

If the 595,192 ignorant colored people and former slaves of Georgia have accumulated property, with title in *fee simple*, as stated above, surely the intelligent immigrant from Europe or the Northern States of the Union can do much better. Can they do as well in the Western States? The above official statement proves that the penniless immigrant who lands in Georgia with a determination to work resolutely can soon acquire a competence.

A glance at the condition of Georgia before the late war, will demonstrate the advantages the State offers as a field for the investment of capital. The wealth of Georgia in 1860, was, in the aggregate, $645,895,237, or nearly $1,100 to each citizen. In 1870, five years after the close of the war, the aggregate was reduced to $268,169,207, being $268 to each inhabitant. Hence, our loss is the emigrant's gain, and a million acres or more, located near railroads or navigable streams, are offered for settlement at a less price than public lands of the U. S. Government.

(The public lands of the United States are divided into two classes: those held at the usual price of $1.25 per acre, and those which lie in sections alternate with railroad lands, and are consequently put at $2.50.) I am authorized to offer land at from $1.00 to $5.00 per acre in quantities to suit purchasers. Land, with improvements, can now be bought for $5.00 per acre, which would have cost the purchaser $50.00 per acre in 1860, and it is intrinsically as valuable now as it was then. Meanwhile the price of land in Georgia is steadily rising in value. There are 35,000 more white voters than colored voters in Georgia.

TERRITORIAL EXTENT, GEOGRAPHICAL POSITION AND CAPACITY FOR POPULATION, WAGES IN GEORGIA AS COMPARED WITH WAGES IN THE NORTHERN STATES.

By Land Measurement there are 37,120,000 acres in Georgia.

The first census was taken in 1790, or 80 years before—Georgia then had only 82,000 inhabitants. In other words, there has been an increase of over a million people in 80 years. In 1790 the State had only 11 counties, of these Wilkes, with 31,000, Richmond, with 11,000, and Chatham, with 10,000, were the largest, and Camden, with 305, was the smallest; and there were only two Representatives in Congress. The State grew rapidly in the next ten years, showing an increase of nearly a hundred per cent. in population, which was the beginning of that growth which now shows 137 counties and nine Members of the United States House of Representatives.

The population of Georgia in 1850, was 906,185 ; in 1860, 1,136,-692; increase, 230.507, or 25 43 per cent.; the population in 1870 was 1,184,109, increase 47,417, or 4 per cent. But for the war the increase from 1860 to 1870 would probably have been 288,720, instead of 47.417, or 60,326 persons per annum actually lost by battle and prevention of increase in population. But the material loss in property has even greater significance when we consider Georgia's present condition. The wealth of the people of Georgia was, in 1850, $335,426,000; in 1860, $645,895,000; increase, $310,469,000, or 90 per cent. In 1870, $268,169,000; decrease, $377,726,000, or 58.5 per cent. At the former rate the increase would have been 90 per cent., $581,305,000, making the wealth of 1870, $1,227,200,000. Actual wealth, 268,169,000; loss, 959,031,000. The loss was more than three times as great as the property left, and the estimate at that in greenbacks, not gold. The above facts are treated *in extenso* in the "Hand Book of Georgia." Thus we have had about seven years, increase in population and twenty-five years' loss of wealth.

Before the war between the States interest was low and material development was rapid. Georgia is a new State, so far as its attractions to immigrants are concerned, in that the price of land has receded from $50 per acre to $5 per acre, while hundreds of thousands of acres may be bought for $1.50 per acre, sometimes less.

In this connection, the following additional quotation is here inserted :

(Special Cor. Boston Advertiser.)

"WOODSIDE, RICHMOND CO., GA. ⎱
"April 2, 1880.   ⎰

"I find, in one of our prominent Northern papers, the following positive assertion : ' *Throughout* (the italics are mine) the South the laws relating to the laboring class are unjust, and if justly framed are not enforced with strict impartiality.' The object of this assertion is apparent ; it is charity to suppose it to be made in ignorance of the facts.

The Bill of Rights in Georgia.

" In this letter I shall confine what I have to say to the State of Georgia. The present Constitution of this State was ratified in 1877; the Convention which framed it was, it must in justice be said, Democratic. The bill of rights guarantees to every citizen natural, civil and religious liberty as large and unclouded as does that of my own State of Massachusetts. Paragraph xxiii., of Section 1, especially provides that ' the social status of the citizen shall never be the subject of legislation.' The laws also give equal protection to all; in effect, however, tending to protect the laborer against the capitalist. They create a lien upon property for taxes, for judgments or decrees of Court, and in favor of laborers, mechanics, contractors and others.

" The wages of laborers cannot be garnished. Article viii of the Constitution provides for a thorough system of common schools, free to all the children of the State, white and black. Without going further into details, which is unnecessary, as the Constitution and laws are accessible to all, it is apparent that the legal provisions for the enfranchisement and protection of labor are abundant.

" I do not know what Southern State is to bear the burden of this charge of legalized injustice to her laboring classes; I do know that it is not to be borne by the State of Georgia.

" The impolicy of unjust and proscriptive laws is better understood in Georgia and the South than it seems to be in California. The laborer is *better paid*, and is, as a class, more contented than his brother laborer in Massachusetts or New York or Pennsylvania.

The Question of Wages.

" Recently, in looking over the accounts of a large plantation, I found the average wage-rate for men and women to be $9.10 per month for full hands. The low wage-rate, when compared to that of the North for the same class of labor, *is more apparent than real.* The full plantation hand who obtains $8 or $9 per month, has, in addition to his stipend, his rations, his house-rent free, his fuel, and usually a small plat of land, upon which he grows with the help of his family, a few necessary vegetables, or perhaps a small bale of the great staple of the South.

" To present the comparison between the plantation hand here and the common laborer at the North more fully, I estimate the plantation hands' wages as follows:

```
12 months' labor, at $9 ......................$108 00
12    "     rations, at 10c. per day ..........  36 50
12    "     rent of cabin...................... 25 00
12    "     fuel...............................  10 00
12    "     rent of house plot ................   5 00
```

Wages of full hand for one year......$184 50

" Our common laborer at the North obtains for similar work:

```
313 days' labor at $1 10 per day.............$344 30
Deducting rent.................... $75 00
Fuel.... .....................  25 00
Food for one person, able-bodied and
    working....................... 100 00    200 00
```

Net wages for the year...............$144 30

"It will thus be seen that the plantation hand realizes in money, or its just equivalent, forty dollars per year more for his labor than his brother laborer in the North. Besides this pecuniary advantage, the negro has a climatic advantage in food and clothing, which the Northern laborer cannot overcome."

As an evidence of the need of population in Georgia, and of the desirability of securing a population of intelligent small proprietors rather than an addition of laborers in the cotton field, the following statistics are offered:

"France has an area of 201,900 English square miles, and a population of 36,102,921, or an average of about 178 to the square mile. This sub-division of the 90,000,000 acres that are cultivated is owned as follows: There are 5,550,000 properties. Of these, the properties averaging 600 acres, number 50,000; those averaging 60 acres, 500,000; while there are 5,000,000 holdings under 6 acres. The peasants are well off, conservative, and contented, though a hundred years ago," says a writer in the *North American Review*, "they were just the reverse."

The population of Georgia is about 20 to the square mile, and a farm of ten acres in Georgia will yield as great annual profit as a similar farm in France will. There are 37,120,000 acres in Georgia. Ireland has an area of 20,819,829 statute acres, or 31,874 square miles. She has a population of 5,411,416. Guernsey (including adjacent islands,) which is entirely divided up into small agricultural holdings, has an area of 19,605 statute acres. and a pupulation of 33,969. Ireland has thus, in rough numbers, a population of one person to every four acres, while Guernsey has a population of nearly two persons to every acre. Yet Guernsey is prosperous, and Ireland is miserable. Guernsey's peasants are proprietors; Ireland's peasants are tenants-at-will. Belgium has 451 persons to the square mile. Italy has 237 inhabitants to the square mile. The Netherlands 179 to the square mile. Switzerland has 175 people to the square mile, but the soil of Switzerland is very equally divided among the people, only about half a million of the total population of 2,669,147 owning no land. From the countries where small holdings or peasant proprietors prevail the emigration is slight. An exception is to be made concerning Germany, for it is not the land system but the *landwher*—a military conscription and a government of force—which impels the mass of German emigrants to free America.

The report of the Congressional Committee on the depression of labor in the United States shows that there are 160,000 Chinamen West of the Sierra Nevadas, besides from 1,500 to 2,500 Chinese women of the vilest character, who are slaves. These people pay less taxes than the Chinese criminal expenses amount to, and send $100,000 a day, $3,000,000 a month, and $40,000,000 a year to China. The money paid them in thirty years, the committee find, has reached $600,000,000; and yet they buy no land, and spend no money, except for a little coarse food, because they get their rice from China. There are no Chinese in Georgia, and in whole counties the white immigrant will have little, if any competition with negro labor, the colored population being very sparse in many parts of the State. The utmost harmony prevails between the two races, however, and the colored people of Georgia are rapidly becoming taxpaying landholders and thrifty farmers.   We do not fear the "exodus," (so-called,) nor do we desire that our negro population should leave the State.   Its broad area of 58,000 square miles, with an average length of 300 miles and an average width of 200 miles, offers better homes to them, and to the poor of other countries, than they can obtain elsewhere.   The average wages paid to mechanics in Georgia is from $1.50 to $3.00 per day.

GEOGRAPHICAL SITUATION OF GEORGIA.

The State of Georgia lies in the Southeastern portion of the United States.

(From the Manual of Georgia).

The nominal divisions of the State are three-fold, to wit: Southern, Middle and Northern Georgia.   These correspond, in the order stated, with the three great natural divisions, viz: the low country, the hill country and the mountain region.

Southern Georgia lies below the line crossing the heads of naviga-tion of the rivers, a portion of which flow into the Atlantic Ocean, and a portion into the Gulf of Mexico.   It is the largest of the three divisions, comprising about 35,000 square miles.   It lies, for the most part, below the level of 300 feet above the sea, the average elevation being about 250 feet.

Middle Georgia lies between the heads of navigation and the eleva-tion of 1,000 or 1,100 feet, the average being 750 feet.   It has an area of 15,000 square miles.

Northern Georgia constitutes the Northern portion of the State, and embraces all the mountains of any note, and much hill country.   It has an area of about 10,000 square miles.   The Eastern half has an average elevation of about 1,500 feet, whilst there are mountain chains that rise to the height of 3,000 feet, and peaks to 4,800 feet. The Western half is much lower, the general elevation being only 750 feet, with mountains up to 2,000 feet.

The average elevation of the surface of the State is 650 feet above the sea.

These three divisions of the State differ in soil and climate, and, to some extent, in productions, as we shall have occasion to note more particularly hereafter, when we come to treat of those several topics.

The mountains of Georgia constitute the Southern terminus of the great Appalachian chain, which, commencing at the mouth of the St. Lawrence, in the North, traverses that portion of America lying East of the Mississippi River, conforming in general direction to the line of coast, until it finally loses itself in Georgia and Alabama, in the South.

Though mountainous, this Northern or Upper Georgia division is, interspersed with rich valleys and hill country, susceptible of cultiva-tion.

The following are the elevations (by U. S. Coast Survey measure-ments,) of prominent mountains in Georgia:

Euota in Towns County............ .... is 4,796 feet high.
Rabun Bald in Rabun County........... " 4,718   "
Blood in Union County............ ..... " 4,468   "
Troy in Habersham County. .... " 4,435   "
Cohutta in Fannin County...... ...... " 4,155   "
Yonah in White County.... .... " 3,168   "
Grassy in Pickens County........ ..... " 3,090   "
Walker's in Lumpkin County........... " 2,614   "
Pine Log in Bartow County............ " 2,247   "
Sawnee in Forsyth County............. " 1,968   "
Kennesaw in Cobb County............. " 1,809   "

Besides these easily recognized mountain ranges, there are other elevated ridges which form the water sheds, separating the drainage areas of the different rivers. The Blue Ridge divides the waters flowing into the Tennessee from those of the Savannah flowing to the Atlantic on the one hand, and those flowing to the Chattahoochee and the Gulf of Mexico on the other.

STONE MOUNTAIN.

Middle Georgia is undulating throughout, without mountains, or level plains to any great extent, and is a very productive portion of the State. With the exception of a narrow belt on the coast, it has been the longest settled. Nearly every acre of it is susceptible of cultivation. The remaining division or Lower Georgia, is, for the most part, a level country, the soil generally light, and the natural forest growth is generally pine—the yellow, long-leaf pine so famous in commerce. This is the finest timber region on the continent, and the rivers and railroads furnish cheap and convenient transportation facilities to Brunswick, Darien and Savannah, whence steamers convey the timber to Northern and foreign ports.

Georgia has a sea-front of about 200 miles, indented with some of the finest harbors on the Atlantic coast. Savannah, Darien, Brunswick and St. Mary's are her principal sea-ports, all of which can communicate by inland navigation through channels running inside of a chain of islands which line the coast throughout its entire length."

Stone Mountain is a remarkable natural curiosity. It is near the Georgia Railroad, sixteen miles from Atlanta. This peak of solid granite is nearly two thousand feet in height and six or seven miles in circumference. The Stone Mountain granite is highly esteemed for building purposes, and is extensively used in Atlanta, Augusta, Macon, and other cities in the State.

BRUNSWICK HARBOR.

According to a trigonometrical survey, made in 1856, under the direction of A. B. Bache, Superintendent of the Coast Survey of the United States, under command of Lieutenant Colonel S. D. Trenchard, the depth of water and rise and fall of the tides at Brunswick, is as follows:

	FEET.
Least depth at low water	18.0
" " " high "	24.8
" " " spring tides	26.2
" " " high "	26.9
Rise of highest tides	8.9
Fall of lowest tides	2.3
Mean rise and fall of tides	6.8
" " " " " Spring tides	8.2

It will be seen from the above official statement that the port of Brunswick can accomodate vessels of the heaviest draft.

The 32d parallel of North latitude passes nearly through the centre of the State.

Georgia, from her geographical relations, is also the natural highway to the teeming products of the great agricultural heart of the country, the Mississippi Valley.

TEMPERANCE STATISTICS.

GEORGIA AND MASSACHUSETTS.

"Wherever a people," says Mr. John Bright, of England, "are not industrious and are not employed, there is the greatest danger of crime and outrage."

Believing that there is less crime and fewer idlers in the South than in any other part of the United States, we offer the following figures gleaned from the report for 1880, of Commissioner Raum, of the United States Internal Revenue Department. That document has the list of retail liquor dealers in the United States. There are 165,850 in the Union. According to Mr. Raum there are in the six New England States the following number of retail liquor dealers.

Massachusetts	6,333
Connecticut	2,372
New Hampshire	826
Vermont	440
Rhode Island	1,275
Maine	694
Total	11,940

Let us take six of the Southern States:

Georgia	2,617
Louisiana	3,559
Mississippi	1,392
North Carolina	1,759
South Carolina	1,272
Texas	3,392
Total	13,981

The census of 1870 gives the New England States 3,487,724 souls. The same census says that the six Southern States above named contained 5,334,812. At this writing the increase in the population of Texas would bring the six Southern States up to about 7,000,000. So it will be seen the New Englanders have two bar-keepers to one in the South. According to proportion of the inhabitants, and were the example of Massachusetts to be followed, Georgia should increase her retailers from 2,372 to over 6,000.

A Northern gentleman, of extensive experience as a traveler, who has passed several winters in Southern Georgia, thus describes it:

"It is pre-eminently the country for men of moderate means to come to. For eight months in the year, the climate is the superb climate of the South of France and Northern Italy. Hundreds of days are like the best days of October and early June in the North, when with the balmy breezes and clear skies, it is simply a pleasure to live.'

TRANSPORTATION.

(From the Manual of Georgia.)

There is a good and safe inland navigation along the Georgia coast, from Savannah to Florida, connecting with the river St. John's, of the latter State, a distance of about 200 miles.

Georgia has 2.396 miles of railroad completed and in operation, or about one mile of road to every 488 inhabitants.

The river Savannah is navigable by steamers the year round from its mouth to Augusta, a distance of about 250 miles, and thence, by flat or "keel" boats, to its confluence with the Broad, about 100 miles further by water.

The Altamaha and its tribuary, the Ocmulgee, are navigable by steamers to Hawkinsville, in Pulaski County, a distance of 340 miles, and will soon be open to Macon, some 60 miles higher up. The Oconee, another tributary of the Altamaha, is open to steamers to the Central Railroad bridge, in Washington County, a distance of 340 miles from Darien.

The Chattahoochee, including the Apalachicola, is navigable from the Gulf of Mexico to Columbus, a distance of 400 miles.

The Flint is navigable 150 miles, to Albany, in Dougherty County, and can be readily opened to a much higher point.

The Coosa runs 40 miles in Georgia, and is open to Greensport, Alabama. Its tribuary, the Oostanaula, is navigable 105 miles above Rome, and work is now progressing to open it 30 miles further.

In addition to these, may be mentioned the Satilla, St. Mary's, Ocholochnee, Ohoopee and Ogeechee Rivers aggregating about 150 miles, making a total of about 2,000 miles of river navigation within the State.

RAILROADS IN GEORGIA.

The following is a list of the railroads in this State, together with the length of each within the State lines:

Western & Atlantic, from Atlanta to Chattanooga, Tenn..138 miles.
Marietta & North Ga., from Marietta to Canton.......... 24 "
Rome Branch, from Kingston to Rome.................. 20 "
Cherokee R. R., from Cartersville to Rockmart.......... 23 "
A. & R. Air-Line, from Atlanta to Charlotte—in Ga......100 "
Northeastern, from Athens to Lula City................. 40 "
Elberton Air-line, from Elberton to Toccoa City......... 51 "
Georgia, from Augusta to Atlanta.......................171 "
Washington Branch, from Barnett to Washington......... 18 "
Athens Branch, from Union Point to Athens............. 39 "
Savannah & Augusta, from Augusta to Millen....... .... 53 "
Georgia Central, from Savannah to Atlanta.............295 "
Sandersville Branch, from Tennille to Sandersville....... 3 "
Eatonton Branch, from Gordon to Eatonton............. 39 "
Thomaston Branch, from Barnesville to Thomaston...... 16 "
Savannah, Griffin and N. Ala., from Griffin to Carrollton.. 60 "
Savannah & Skidaway, from Savannah to Isle of Hope.... 9 "
Montgomery Branch, from Isle of Hope to Montgomery.. 4 "
Savannah, Florida and Western, from Savannah to Bain-
    bridge.---237 "
Live Oak Branch, from Lawton to Live Oak, Fla.......... 48 "
Albany Branch, from Thomasville to Albany............. 60 "
Macon & Augusta, from Macon to Camak, Ga. R. R..... 74 "
Macon & Brunswick, from Macon to Brunswick..........186 "
Hawkinsville Branch, from Cochran to Hawkinsville...... 10 "
Brunswick & Albany, from Brunswick to Albany.........172 "
Southwestern, from Macon to Eufaula, Ala.............140 "
Muscogee Branch, from Fort Valley to Columbus........ 71 "
Perry Branch, from Fort Valley to Perry................. 11 "
Albany Branch, from Smithville to Albany.............. 23½ "
Arlington Branch, from Albany to Arlington............. 35¾ "
Fort Gaines Branch, from Cuthbert to Fort Gaines....... 22 "
North & South. from Columbus to Kingston............. 20 "
Selma, Rome & Dalton, from Dalton to Selma, Ala.—in Ga. 67 "
Atlanta & West Point, from Atlanta to West Point......... 86¾ "
East Tennessee, from Dalton to Bristol, Va.—in Ga...... 18 "
Chattanooga & Alabama, from Chattanooga to Selma,
    Ala.—in Georgia................................... 25 "
Dodge's R. R., from Eastman, Dodge Co., to Ocmulgee R.
    completed....... .... 18 "

A railroad will be constructed during the next two years between Macon and Atlanta, and it is probable that another will be built from Rome to Chattanooga. The North and South Railroad is being rapidly extended, and will probably be continued until it reaches Rome *via* La Grange.

Senator Brown, of Georgia, thus described, in a speech delivered in United States Senate, June 1st, 1880, the value of Savannah, Georgia, as a National port, and the transportation advantages of Georgia:

" The port of Savannah is the third cotton port in the union. It has a large and growing commerce. During the business season more than one foreign steamer per day clears from the port. The railroad connections that have been recently made and the extensions of lines that have been completed within a short time past, are such that the line leading to Savannah is looked to in the future as one of the great competing lines for the business between the West and the Eastern cities. There is a railroad line now pretty well consolidated under the control of the Louisville and Nashville railroad company from the city of St. Louis, Missouri, to the city of Chattanooga, Tennessee. That corporation controls that line. In connection with that is the Western and Atlantic railroad of Georgia. That road is an open highway under its charter to continue its use for all the roads that approach it, having a web of five roads at the Chattanooga end and four at the Atlanta end. That, then, extends the line from Chattanooga to Atlanta. Then we have the Central railroad to Savannah.

AVENUE IN BONAVENTURE CEMETERY, SAVANNAH.

" Then there is a line of steamers controlled mainly by the Central railroad that plies between the port of Savannah and New York. They have already five steamers on the line. They are building another splendid steamer for it. They intend to increase the number until they have a daily steamer between Savannah and New York.

"As I am informed, a careful estimate shows the line from St. Louis to New York by way of Savannah, on principles recognized among railroad and steamship men in prorating, is seventy miles shorter than the line by way of the Pennsylvania railroad.

The port of Savannah, then, is a national port. The West is as much interested as the South is in the port. The roads I have mentioned extend from the State of Missouri through the States of Illinois, Indiana, Kentucky, Tennessee, a portion of Alabama and through Georgia to the coast. There are seven states that are directly interested in a line that proposes to open a way for the productions of the great West to the markets of the world through the South, by the port of Savannah."

A straight line from St. Louis, or the mouth of the Ohio, shows that the Atlantic coast of Georgia is much nearer and more accessible to the commerce of the West than that any other State in the Union.

PASSENGER DEPOT AND ENVIRONS, ATLANTA.

No part of Georgia has, of late years, increased more rapidly in wealth and population than the Northern section of the State. To this part of Georgia belongs Atlanta, the capital, a city remarkable for its rapid growth, as well as for the enterprise and public spirit of its people. In 1860 the population was 10,000. In 1864 Atlanta was almost totally destroyed by Gen. Sherman. The old citizens returned to the ashes of their former homes in 1865, and began to rebuild the city.

The following table will show the growth of Atlanta and its neighbors for the past ten years.

	1870.	1880.	Increase.
Atlanta	21,789	*40,000	18,211
Marietta	1,888	2,575	687
Gainesville	472	1,942	1,470
West End	621	776	155

*This is an approximation only. *The Chicago Tribune* estimating the population of Atlanta at 45,000, states that the ratio of increase is larger in Atlanta than in any city in the United States, except Denver, Col., and

PERSONAL RIGHTS.

No greater personal rights are granted to immigrants in any State of the Union than in Georgia. Every man is allowed to think, and speak, and vote as he pleases. If he is harmed in person, property or character, for exercising this inestimable privilege, the law will zealously enforce his rights and is a sure remedy. The law creates a lien upon property in favor of laborers, mechanics and landlords. There is no such thing as imprisonment for debt, and a reasonable amount of property is exempt from seizure or sale on account of debt. Religious freedom prevails, and there is no connection between Church and State. No religious qualification is necessary for office, or necessary to constitute a voter, the only condition being that the man shall be 21 years old, and shall have complied with the naturalization laws of the United States. The people choose their own officers and one man's ballot is as good and as powerful as another's. Any citizen of the State has a right to hold any office in the State, for he is no longer a "foreigner" after he becomes a citizen of Georgia. The homestead exemption will prevent any creditor from taking away the home of your family.

HOMESTEAD EXEMPTION.

The Constitution of Georgia exempts from levy and sale, by virtue of any legal process whatever, (except in the cases named below,) of the property of every head of a family, or guardian, or trustee of a family of minor children, or every aged or infirm person, or person having the care and support of dependent females of any age, real or personal estate, or both, to the value, in the aggregate, of *sixteen hundred dollars*. Said property, however, is liable to levy and sale for taxes, for the purchase money of the same, for labor done thereon, for material furnished therefor, or for the removal of incumbrances thereon. The exemption includes not only the property itself, but all improvements made thereon after it is set aside. A mortgage of property by the father during his lifetime, cannot, after his death, deprive his minor children of a homestead, or exemption right in the mortgaged premises.

LEGAL PROVISIONS OF GENERAL INTEREST.

No person shall be deprived of life, liberty, or property, except by due process of law.

Every person has the right to prosecute or defend his own cause in any of the courts, in person, by attorney, or both.

Every person charged with an offense against the laws of this State, shall have the privilege and benefit of counsel; shall be furnished, on demand, with a copy of the accusation, and a list of the witnesses on whose testimony the charge is founded; shall have compulsory pro-

Minneapolis, Minn. I quote as follows from the Tribune: "The cities showing the largest ratio or increase are Atlanta, Ga., 106 per cent.; Denver, Col., 614 per cent.; Minneapolis, Minn., 244 per cent.; Milwaukee, Wis., 92 per cent.; St. Paul, Minn., 100 per cent.; Waterbury, Conn., 103 per cent.; Meriden, Conn., 80 per cent; and Camden, N. J., 84 per cent. It will be noticed that there has been a very surprising growth in manufacturing cities.

cess to obtain the testimony of his own witnesses; shall be confronted by the witnesses testifying against him, and shall have a public and speedy trial by an impartial jury.

The rate of taxation in Georgia, for State purposes, varies from year to year, according to the wants of the government from 70 cents to 100 cents on each $100 worth of property. The several counties are authorized, in addition, to levy a tax for county purposes, not to exceed fifty per cent. on the amount of State tax levied for the same year.

Perfect freedom to worship God according to the dictates of his own conscience, is guaranteed to every citizen.

No inhabitant of this State shall be molested in person or property, or prohibited from holding any public office or trust, on account of his religious belief.

No law shall ever be passed to curtail, or restrain the liberty of speech or of the press.

The right of the people to be secure in their persons, houses, papers and effects against unreasonable searches and seizures, shall not be violated; and no warrant shall issue except upon probable cause, supported by oath or affirmation, particularly describing the place or places to be searched, and the person or things to be seized.

The social status of the citizen shall never be the subject of legislation.

There shall be no imprisonment for debt.

The right of the people peaceably to assemble, and, by petition or remonstrance, apply to the government for a redress of their grievances, shall not be denied.

All citizens of the United States, resident in this State are to be considered citizens of this State, and the Legislature shall make all necessary laws for their protection as such.

No conviction shall work corruption of blood or forfeiture of estate.

Private property shall not be taken, nor damaged for public purposes without just and adequate compensation to the owner.

No *ex post facto* law, or law impairing the obligation of contracts, shall be passed.

No total divorce shall be granted, except on the concurrent verdicts of two juries, at different terms of the Court.

Cases respecting titles to land shall be tried in the county where the land lies. All other civil cases shall be tried in the county where the defendant resides, and all criminal cases in the county where the crime was committed, except cases in the Superior Court, where the Judge is satisfied that an impartial jury cannot be obtained in the county.

Grand jurors are drawn from the body of the people, and must be experienced, intelligent and upright men. Traverse jurors are drawn in the same way, and must be intelligent and upright men.

Titles to land can be passed only by will or deed in writing duly executed.

The entailment of estates is prohibited by law. Gifts or grants in tail convey an absolute title.

In making his will, a testator may do what he chooses with his property, except that he cannot prejudice his creditors; and the law considers his wife so far a creditor that he cannot deprive her of dower, except with her consent.

A wife, notwithstanding marriage, continues to be the legal owner of the property she possessed at the time of marriage, and of any that may accrue to her by gift, bequest, or her own acquisition, after marriage.

Upon the death of an intestate, his widow may elect to take a dower or one-third interest for life, in the lands of her deceased husband, and share and share alike with the children in the personal property; or, she may relinquish her right of dower, and take a child's part, share and share alike, in all the property, to be her own, absolutely.

STATE CHARITABLE INSTITUTIONS.

(*From the Manual of Georgia.*)

THE GEORGIA ACADEMY FOR THE BLIND is located at Macon, Bibb County, and is supported almost entirely by the State.   Pupils of both sexes, between the ages of 7 and 25 years, are admitted, though males over 25 are received for instruction in the various trades.   The pupils are taught all the elementary branches of an English education, together with the Holy Scriptures, history and music.   They are also instructed in such mechanical trades as can be

ORPHAN'S HOME, AUGUSTA.

imparted to the sightless.   The blind are thus redeemed from ignorance through this public charity, and taught to be useful, and even
self-supporting.   The school is conducted by a Principal, two Profes-
sors, three assistants, and a Master of Workshops.

DEAF AND DUMB INSTITUTE.—This school is located at Cave
Spring, in Floyd County, one of the most romantic and delightful
sections of the State.   It is also supported by the State, and annually turns out quite a number of this truly unfortunate class, educated
sufficiently to enable them to enjoy free intercourse with others,
and with occupations by which, with proper industry, they have no
difficulty in earning a support.   The most approved system of instruction is adopted in this institution, as well as in the Academy for
the Blind.   Besides the Principal, there are four male teachers, one
female teacher and a matron.

LUNATIC ASYLUM.—The State of Georgia, many years ago,
erected, at a heavy cost, near Milledgeville, then the seat of government, an Asylum for the care of lunatics, and, by means of liberal
appropriations, has enlarged the charity from year to year until the
inmates have come to number 845, of whom there are whites, 710—
376 male, and 334 female.   The blacks are separately provided for
and number 135, of whom 69 are males, and 66 females.   The Asylum
is both a home and a hospital.   Under skillful physicians and nurses,
the patients receive the best of attention, and large numbers are annually restored to health of mind and body.   About ten years ago,
the Legislature passed an act setting aside the Okefinokee Swamp,
containing about 500,000 acres of land—decided to be reclaimable at
a small cost compared with its value for timber and agricultural purposes, as a permanent endowment for a State Orphans' Home.

BENEVOLENT SOCIETIES.

The FREE AND ACCEPTED MASONS, the most ancient of orders
and by far the most numerous in the State, has over 280 lodges and
about 15,500 members.   There are also eight chartered Command-
eries, with 320 members.

The INDEPENDENT ORDER OF ODD FELLOWS has 48 lodges, and
about 2,000 members.

The KNIGHTS OF PYTHIAS, KNIGHTS OF HONOR and SONS OF
MALTA have each lodges in all of the cities and some of the smaller
towns of the State, but we have failed to obtain definite information
regarding them.

The INDEPENDENT ORDER OF GOOD TEMPLARS, a benevolent
order, with a pledge of total abstinence from intoxicating drinks, has
3co working lodges and a membership numbering 10,000.

Besides these, nearly all the churches have relief societies, and
benevolent associations exist in all the cities of the State.   A meri-
torious claim to charity, or temporary aid, is seldom disregarded in
Georgia.

Georgia has 2,396 miles of railway transportation, 2,000 miles of
river transportation, public schools free to all the children of the
State, 213,000 spindles in operation, a climate unexcelled, and a

"Bill of Rights" that guarantees as much civil, religious, political, and social liberty as any American State can offer. Immigrants are cordially received, there is no class society, and integrity, industry and sobriety will admit the new comer into any society which his education would enable him to enter at the North or West.

WESLEYAN FEMALE COLLEGE.

Macon, the fourth city of Georgia, is situated on both sides of the Ocmulgee River, at the head of steamboat navigation, and is also in the middle section of the State. It is surrounded by a productive country, and is connected by rail with the cities of Atlanta, Columbus, Augusta, Savannah, Albany, and Brunswick. The first lots were sold in 1823. It is now a thriving and beautiful city. It has an extensive trade, large foundries, a cotton-factory, flouring mills, and planing-mills. Its yearly receipts of cotton are seventy-five thousand bales.

Macon might appropriately be called "the city of colleges." The Wesleyan Female College, belonging to the Southern Methodist Church, ranks among the best institutions of the kind in the Union. It has the honor of being the first college in the United States to confer diplomas upon females.

EDUCATION IN GEORGIA.

Georgia lost by the war $377,726,000.

The population of Georgia, as stated in the census of 1870, the last taken, numbers 1,184,109. Of these 638,926 are white, and 595,192 black and mulatto, who became free citizens in 1865.

---

* The census of Macon in 1860 was 7,247; in 1870, 10,810; an increase of fifty per cent. The present census is not yet complete, but a hasty glance shows that within the corporate limits the increase in the last ten years has been equally as great. The city is full; there are not six houses in its corporate limits for rent. The postal business of Macon is greater than that of any city (Atlanta excepted) in North and South Carolina, Georgia, Florida and Alabama. The wholesale dry goods and shoe trade of this city is larger than that of any city in the State.

It will thus be seen that the conditions which rendered it advisable to establish free public schools for the whole population, without prejudice as to color, race or previous condition, were very different from those which existed in the North and West when free public schools were established there. In 1795 the lands in Ohio known as the "Western Reserve," belonged to Connecticut, and were sold in that year for $1,200,000. This sum was consecrated to the support of free schools in Connecticut. Massachusetts held wild lands in the colony of Maine, and thus provided a fund for the support of free schools. In the Western States generally, the 16th section of every township was granted for school purposes by the national government. The lands thus granted amounted to 68,000,000 acres, and were sold for $60,000,000. Georgia has had none of these advantages. The common school funds in the various States consist of grants of lands made by the general government supplemented by poll and property taxation. In Georgia we have to depend on taxation alone. Knowing that the records of crime and pauperism prove that the proportion of criminals from the illiterate classes is ten-fold greater than from those having the benefits of education, and the proportion of paupers among the former is sixteen times greater than among those of the educated classes, the State of Georgia has done all that in its impoverished condition after the late war it could do, but will not stop in the good work until every child in Georgia is at liberty to receive education at the expense of the commonwealth.

The following table will demonstrate the progress already made, as found in the annual report of the State School Commissioner published on November 6th, 1878.

"There have been enrolled in the schools in the successive years since the beginning of the work as follows: In 1871, white pupils, 42,914; colored, 6,664; total, 49,578; in 1873, white, 63,922; colored, 19,755; total, 83,677; increase over the attendance of 1871, 34,099; in 1874, white, 93,167; colored, 42,374; total, 135,541; increase over the attendance of 1873, 51,864; in 1875, white, 105,990; colored, 50,358; total, 156,394; increase over the attendance of 1874, 20,808; in 1876, white, 121,418; colored, 57,987; total, 179.405; increase over the attendance of 1875, 23,011; in 1877, white, 128,296; colored, 62,330; total, 190,626; increase over the attendance of 1876, 11,221.

The Commissioner further says:

"In 1874, the number of persons between ten and eighteen years old unable to read, was reported as follows: white, 26,552; colored, 79,692; total, 106,244. The same figures as repoated this year, are as follows: white, 22,223; colored, 63,307; total, 85,630. Notwithstanding the great increase in school population reported, the decrease in the number of illiterates is as follows: white, 4,229; colored, 16,385; total, 20,614. These figures show that our effort towards the extinguishment of illiteracy, notwithstanding the meagerness of the means put at our disposal, have not been without results, and especially do they show that our colored citizens are not disposed 'to despise the day of small things,' as to educational privileges."

Yet it will be seen that Georgia has not waited for Congressional aid, but, in all that goes to make the laboring classes comfortable and prosperous, challenges comparison with any Northern State of similar population. The colored people have developed a great desire to get an education, and remarkable aptitude in acquiring it. The testimony of the board of visitors annually appointed by the governor declares that the capacity displayed by the colored pupils is in every respect equal to that displayed by white pupils, within the same field and under the same limitations. This is peculiarly gratifying to the white citizens of Georgia, who have voluntarily done all in their power to advance the material, moral and educational interests of this hitherto dependent race. Last year a public school was established in every militia district in the State. Nearly five thousand were in operation, and two hundred and ten thousand children were given schooling. This year the fund will be supplemented by the tax on retail liquor establishments, and by this and other means the commissioner hopes to extend the school term in every portion of the State.

Mr. Hoar, of Massachusetts, has introduced a bill in Congress, providing that the receipts from the public lands and the patent office, and the income from the public lands shall be appropriated as a school fund, the most of which, during the first ten years, is to be spent in the Southern States.

SOILS OF GEORGIA.

The traveler in the great West must be impressed with the uniformity of the landscape. Almost as flat as Holland, the vast prairie seems endless as sea or sky, and the eye of the immigrant is delighted with the luxuriant crops which indicate a rich soil. He forgets that the poor farmer is in the grasp of the octopus-like railway corporations that control transportation and discriminate in freight charges against the poor producer and in favor of the rich landlord. He forgets that *hay is used for fuel*, and that the cost of transportation to market is so great frequently, that the luxuriant crops of corn made cannot be sold at a profit. He forgets that the civilized world is the competitor of the wheat grower, and that it is estimated that the wheat crop of the present year will exceed the demand for export by 100,000,000 bushels. He forgets that the railroads have so far overcome competition, that in future Western grain must be expected to pay much more for carriage to the seaboard than heretofore, while ocean freights are not likely to rule so low as during the last few years of commercial

SCENE IN NORTHERN GEORGIA.

depression. One of the best authorities in the United States, the *New York Commercial Bulletin*, thus warns him:

"Our agriculture is taking its seat at such an immense distance from the seaboard, that either the railroads or the farmers must suffer serious deductions for transportation, in order to compete with European farmers. And, what is of more immediate importance, our grain crops have reached such a state of over-supply, that when Europe regains average seasons, we cannot hope to escape a depression in the prices of grain that will compel a comparative contraction of that branch of industry."

CAPACITY OF GEORGIA SOIL UNDER HIGH CULTURE.

(*From the Manual of Georgia.*)

We proceed to give the results of a number of experiments in the cultivation of those products, in each of those divisions, conducted with proper preparation and fertilization—such as are given in the more densely settled portions of the world. As but little is accomplished by inadequate means in any department of human industry, the actual producing capacity of a country can only be tested by the results of *judicious* culture. The crops to which we shall refer, were reported to the various State and county fairs within the past few years, and both the culture and its results were verified by the affidavits of disinterested parties.

In 1873, Mr. R. H. Hardaway produced on upland, in Thomas County, (Lower Georgia), 119 bushels of Indian corn, or maize, on one acre, which yielded a net profit of $77.17.

In the same county, the same year, Mr. E. T. Davis produced 56½ bushels of rust-proof oats per acre. After the oats were harvested, he planted the same land in cotton, and in the Fall gathered 800 pounds of seed cotton.

Mr. John J. Parker, of the same county, produced, in 1874, on one acre, 694½ gallons of cane syrup, at a cost of $77.50. The syrup, at 75 cents per gallon, the market price, brought $520.87—net profit from one acre, $443.37.

In 1874, Mr. Wiley W. Groover, of Brooks County, (Lower Georgia) produced, with two horses, on a farm of 126½ acres, without the aid of commercial fertilizers, cotton, corn, oats, peas, sugar cane and potatoes, to the value of $3,258.25. The total cost of production was $1,045.00, leaving net proceeds of crop, $2,213.25. The stock raised on the farm was not counted.

Joseph Hodges, of the same county, produced, on one acre, 2,700 pounds of seed cotton; Wm. Borden, 600 gallons of syrup; J. Bower, 500 bushels of sweet potatoes; J. O. Morton, 75 bushels oats. Mr. T. W. Jones made 12 barrels, or 480 gallons of syrup on one acre, and saved enough cane for seed.

In Bulloch County, (Lower Georgia) 3,500 pounds seed cotton were produced by Samuel Groover, and in the same county 21 barrels sugar at one time, and 700 gallons syrup at another, per acre.

In Clay County, Mr. —— Hodge produced from one acre, a few years ago, 4,500 pounds of seed cotton.

Mr. J. R. Respass, of Schley County, gathered the present year (1878) a little upwards of 500 bushels of oats from five acres.

Mr. J. R. Respass, of Schley county, (Lower Georgia) in 1877, by the use of fertilizers, grew on five acres of naturally poor land, 15,000 pounds of seed cotton, which netted him when sold $66.02 per acre.

Mr. H. T. Peeples, of Berrien County, reports to this Department a crop of 800 bushels of sweet potatoes grown on one acre of pine land.

In 1876, Mr. G. J. Drake, of Spalding County, (Middle Georgia), produced 74 bushels of corn on one acre of land.

Mr. John Bonner, of Carroll County, made three bales of cotton (500 pounds each) on one acre. Mr. R. H. Springer, of the same county, produced nine bales from five acres, without manures, and ninety-four bales from one hundred acres by the use of fertilizers.

In 1873, Mr. S. W. Leak, of the same county, produced on one acre 40¼ bushels of wheat, worth $80.50; cost, $14.50; net profit, $66.

In Wilkes County 123 bushels corn was produced on one acre of bottom land; also 42 bushels Irish potatoes on one-tenth acre, the second crop same year on same land; the first crop was very fine, but not so good.

Mr. J. F. Madden, of the same county, produced in 1876, on one acre, 137 bushels of oats.

Mr. T. C. Warthen, of Washington County, (on the line of Middle and Lower Georgia,) produced in 1873, on 1.1125 acres, 6,917 pounds of seed cotton, equivalent to five bales of 461 pounds each, worth at 17½ cents per pound, the average price of that year, $403.37. ˙ The cost of culture was $148.58; net profit, $254.79 for a very small fraction over one acre.

Dr. Wm. Jones, of Burke County, produced 480 gallons syrup on one acre. Wesley Jones, of the same county, produced three bales of cotton, 500 pounds each per acre. Jas. J. Davis, in same county, made in 1877, with two mules, thirty-four bales of cotton, 500 pounds each, 600 bushels corn, and 300 bushels oats. Wm. C. Palmer, of same county, made in 1877, with one mule, twenty-five bales of cotton, 500 pounds each, and a fair crop of corn. Henry Miller of same county, produced in 1877, sixty-five bushels corn per acre, first year, on reclaimed swamp, without manures.

Mr. R. M. Brooks, of Pike County, (Middle Georgia), produced in in 1873, on five acres of bottom land, 500 bushels of rice. The total cost was $75; net profit $300.

Mr. R. B. Baxter, of Hancock County, (Middle Georgia), in 1872, harvested at the first cutting, first year's crop, 4,862 pounds of dry clover hay per acre.

Mr. A. J. Preston, of Crawford County, gathered from one acre of Flint River bottom 4,000 pounds seed cotton, and from another on same place, 115 bushels corn.

Dr. T. P. Janes, of Greene County, (Middle Georgia), produced in 1871, five tons of clover hay per acre in one season, at two cuttings.

Mr. Patrick Long, of Bibb County, (on the line of Middle and Lower Georgia) harvested from one acre of land, from which he had gathered a crop of cabbages in June of the same year, 8,646 pounds of native crab-grass hay.

Mr. S. W. Leak, in Spalding County, (Middle Georgia), gathered in the Fall of 1873, from one acre, from which he had harvested forty bushels of wheat in June, 10,726 pounds of pea-vine hay. Net profit from wheat, $66; from pea-vine hay, $233.08, making in one year from a single acre, a net profit of $299.08.

Mr. William Smith, of Coweta county (Middle Georgia,) produced 2,200 pounds seed cotton per acre on ten acres.

Mr. Edward Camp, of the same county, produced 1,000 bushels oats from ten acres.

Mr. J. T. Manley, of Spalding county, (Middle Georgia,) produced 115 bushels of oats from one acre.

Mr. S. W. Bloodworth, of the same county, gathered in 1870, 137 bushels of corn from one acre.

Mr. L. B. Willis, in Greene county, (Middle Georgia) in June, 1873, from one acre and a third, harvested twenty bushels of wheat, and the following October, 27,130 pounds of corn forage. From the forage alone he received a profit of $159.22 per acre.

Dr. W. Moody, of the same county, harvested, at one cutting, from one acre of river bottom, in 1874, 13,953 pounds of Bermuda grass hay; cost, $12.87, value of hay, $209.29, net profit, $196.42.

Mr. J. R. Winters, of Cobb county (Upper Georgia,) produced in 1873, from 1.15 acres, 6,575 pounds of dry clover hay at the first cutting of the second year's crop.

Mr. T. H. Moore, of same county, produced on one acre 105 bushels of corn, while Mr. Jeremiah Daniel produced 125 bushels.

Mr. R. Peters, Jr., of Gordon county, (Upper Georgia,) harvested in 1874, from three acres of lucerne, four years old, fourteen tons and 200 pounds of hay, or 9,400 pounds per acre.

Capt. C. W. Howard produced, on Lookout Mountain, in Walker county (Upper Georgia,) in 1874, on one acre of unmanured land, which cost him twenty-five cents per acre, with one hoeing and plowing, 108½ bushels of Irish potatoes, which he sold in Atlanta at a net profit of $97.25. On land manured and better prepared and worked, double that quantity could be produced.

Mr. Thomas Smith, of Cherokee county, produced 104 bushels of corn from one acre.

Mr. John Dyer, of Bibb county, produced in 1873, from one acre, at a cost of $8 00, 398.7 bushels of sweet potatoes, which he sold at a net profit of 290.92.

Mr. Haddon P. Redding, of Fulton county, in 1877, produced from one acre, 400 bushels of St. Domingo yam potatoes, which he readily sold in Atlanta at an average price of $1.00 per bushel.

These instances of production are exceptional, and far beyond the usual results of farming in our State; but they serve to show the capacity of our soil when properly fertilized, and cultivated with intelligence under the guidance of science. It will not be denied, however, that what the parties named have accomplished on a limited scale, may be done by others on still larger areas, and with corresponding results.

THE AVERAGE YIELD AND PRODUCTIONS OF GEORGIA.
NORTH-EAST GEORGIA.

To the robust, no climate can be more bracing or delightful than the beautiful mountain region of North-east Georgia. Mount Yonah, the Cohutta range, the Currahee and others, with peaks attaining sometimes an altitude of 5,000 feet, form a landscape resembling that of the "Saxon, Switzerland." From the summit of Tray mountain hundreds of peaks are visible.

In North-east Georgia,

" The clay, or sub-soil, is usually found from four to six inches below the surface on uplands, from one to two feet in the valleys, and from two to six feet in river bottoms. The original forest growth is chiefly red, black, post and white oaks; chestnut, black-jack, hickory, short-leaf and spruce-pine, cedar, dogwood, black-gum, walnut, with poplar, ash, elm, sycamore, birch, sweet gum and white oak on the lowlands. This is the great auriferous region of the State, the net yield of gold being equal to that of any section of the Union, California not excepted. Copper, lead, magnetic iron ore, mica, asbestos marble, ruby, serpentine, corundum, are also found in considerable quantities, and may be mined with profit.

" The lands are generally rich and productive, the yield depending wholly on the skill used in their cultivation.

" This division embraces nineteen counties, stretching from the Savannah and Tugalo rivers in the east, to the Cohutta range of mountains in the West. It is that part of the State which possesses the greatest elevation, the average being 1,500 feet above the level of the sea, while there are peaks which rise to an elevation of nearly 5,000 feet."

This region comprises the following counties :

Banks,	Dawson,	Fannin,	Forsyth,
Franklin,	Gilmer,	Gwinnett,	Habersham,
Hall,	Hart,	Jackson,	Lumpkin,
Madison,	Milton,	Pickens,	Rabun.
Towns,	Union,	White,	

"The staple field products are Indian corn, wheat, oats, rye, barley, clover, the various grasses, and sorghum cane, while in the Southern portion of the division cotton is grown to a considerable extent. The average yield per acre under fair cultivation, is : corn, 20 bushels ; wheat, 15 bushels ; oats, 25 bushels ; rye, 8 bushels ; barley, 25 bushels ; hay, from 2 to 3 tons ; sorghum syrup, 75 gallons ; cotton, 400 pounds in the seed. Under high culture, two, three and sometimes four times this production is realized. Tobacco, buckwheat and German millet can also be grown with great success. The planting and harvest times of the division are as follows : corn, 15th March to 15th May, gathered in fall months ; wheat and other small grain sowed in October, harvested in June and July ; cotton planted 15th April to 15th May, gathered in fall months ; sorghum, planted in April, cut in August. A very large proportion of the laborers, both farm and mine, are white.

" The fruits best adapted to the section are, the apple, cherry, pear, grape, plum, in all its varieties, peach, gooseberry, raspberry, straw- berry—the last named producing equally well in all parts of the State with like cultivation. Almost every variety of vegetable attains to great perfection.

" The climate is unsurpassed on the continent for comfort and salubrity during nine months of the year. The mean temperature in Summer is 70°, Fahrenheit ; in Winter, 35° ; highest temperature, 90° ; lowest, 8°—periods of greater heat and cold being exceptional. Snow falls usually from two to three times during the winter season, especially in the northernmost counties, to a depth, varying from two to six inches. In the Southern tier of counties, there are occasional winters without a fall of snow.

" Springs and running streams abound in all parts of the district ; water powers unsurpassed ; spring and well water freestone, and not excelled in any country. Mineral springs—sulphur or chalybeate— abound in nearly all the counties of the district."

Northwest Georgia, particularly the lime-stone soils, is peculiarly adapted to Red Clover and the English Grasses, and is more fertile generally than Northeast Georgia, and there the cheapest charcoal pig iron in the United States is made.

There is much picturesque scenery in Northwest Georgia. The view from High Point and Point Lookout, on Lookout Mountain, is magnificent. Seven States can be seen from these points.

The soils are calcareous and argillaceous ; clay, red and yellow. In all other respects our description of the natural conditions and capabilities of Northeast Georgia will apply to this division, with the single exception of temperature, the difference in elevation being accompanied by the usual variations of heat and cold. The productions are, in all respects, the same.

In one or two respects this division enjoys peculiar advantages over its eastern neighbor. It has not only a larger area of tillable land, but a much greater proportion of valley and bottom. Its facilities for transportation are also greater, the Western and Atlantic Railroad traversing its centre from the northern to the southern boundary, while tributary roads supply a good portion of the country to the right and left of the main line.

"Its average elevation above the sea is only 750 feet, or about 50 per cent. less than that of Northeast Georgia. The characteristic minerals are limestone, slate, iron ores, coal, manganese, sandstone, baryta, some gold, all of which, except the last, are found in great quantities. Several valuable veins and gravelly deposits of gold have been developed and worked, with handsome returns.

"Immense coal beds lie in the Northwestern Counties of this division, to wit: Dade, Walker and Chattooga. The supply seems to be inexhaustible; the mines are reached by railroads which connect with main trunks, and, in the immediate vicinity, are immense deposits of best iron ore.

"The following counties are included in Northwest Georgia:

Bartow,	Catoosa,	Chattooga,	Cherokee,
Cobb,	Dade,	Floyd,	Gordon,
Haralson,	Murray.	Paulding,	Polk,
Walker,	Whitfield.		

"Northwest Georgia extends from the Cohutta Mountains and Chattahcochee Ridge to the Eastern boundary of Alabama."

MIDDLE GEORGIA.

The original forest growth consists of red, post, Spanish and white oaks, and black-jack, hickory, short-leaf pine, with some long-leaf on its Southern border; poplar, dog-wood, elm, chestnut, maple, beech, birch, ash, black locust, sweet and black gums, walnut and some cedar. This division has three varieties of soil—red or clay, gray and gravelly, and light and sandy, the last named being limited in extent and confined to the long-leaf pine localities on the Southern border. The two former possess great productiveness and durability. After the coast country they were the first settled, and Middle Georgia has continued to be the most populous divison of the State. While the lowlands are of the best quality, the uplands are unsurpassed in fertility and luxuriance of forest growth.

This division embraces thirty-nine counties, and has an area of about 15,000 square miles. It extends across the State from the Savannah River in the East, to the Chattahoochee River in the West. Its Southern border may be described with tolerable accuracy by a line from Augusta through Macon to Columbus. It is marked by the head of navigation of the principal rivers. The Northern border may be described by a line running through Athens and Atlanta. It is about one hundred miles in width. Its average elevation is 750 feet. The entire region is metamorphic; its rocks, granite, gneiss, mica, quartzites, hydro-mica schist, with some limestone and soapstone. These rocks all extend from the Northeast to the Southwest, and are crossed frequently at right angles by trap dykes. Its chief minerals are gold, copper, lead, asbestus, graphite, chromic iron, serpentine and soapstone. Gold is found in districts wide apart, and has been worked with satisfactory profit in a few localities, more especially in McDuffie, Lincoln, Wilkes and Carroll. Asbestus is also mined to some extent.

" This section embraces thirty-nine counties, as follows:

Baldwin,	Douglas,	Jones,	Putnam,
Bibb,	Elbert,	Lincoln,	Rockdale,
Butts,	Fayette,	McDuffie,	Spalding,
Campbell,	Fulton,	Meriwether,	Talbot,
Carroll,	Greene,	Monroe,	Taliaferro,
Clarke,	Hancock,	Morgan,	Troup,
Clayton,	Harris,	Newton,	Upson,
Columbia,	Heard,	Oconee,	Walton,
Coweta,	Henry,	Oglethorpe,	Warren,
DeKalb,	Jasper,	Pike,	Wilkes,

"The staple field products are cotton, corn, oats and wheat, while all the grains and grasses, and even tobacco, may be grown successfully. The average yields with ordinary culture, are: Cotton, 550 pounds in seed per acre; corn, 12 bushels; wheat, 8 bushels; oats, 25 bushels; barley, 30 bushels; rye, 8 bushels; sweet potatoes, 100 bushels; field peas with corn, 5 bushels. Ground peas, chufas, pumpkins, and, indeed, almost every field product, are successfully cultivated. Very many farmers double the above averages year after year, whilst under high culture the product is multiplied four or five times, as will be seen in the chapter on that subject.

" The planting and harvest periods of leading products are: Cotton, April, September to December; corn, March, October; wheat, October and November, May and early June; other Fall grains harvested same time; those sowed in February and March harvested in June. The fruits to which the section is best adapted are the peach, fig, apple, pear, strawberry, raspberry, melons of all kinds. The peach attains here and in Southwest Georgia, its greatest perfection, and immense quantities are raised for export, both in their natural and dried state; the same may be said of the apple and blackberry, though the latter is a spontaneous growth, and yields abundantly in a wild state. Almost every other variety of fruit known in the Southern States thrives well in this division. The table vegetables are all grown successfully, the hardier varieties the year round. The climate is a happy medium between those of Southern and Northern Georgia, and, in healthfulness, equal to that of any part of the world. There is much uniformity of temperature, sudden rises and falls occurring but rarely. The mean annual temperature is 60° to 64°. Snow falls about once in three years, the depth varying from 1½ to 4 inches. Every portion of the division abounds in running streams, while the spring and well waters are excellent. The difference in elevation between the Northern and the Southern portions of the division being from 650 to 700 feet, the water powers are probably unequaled by those of any similar area on the continent. It would be difficult to fix a limit to its manufacturing facilities in this respect."

EAST GEORGIA.

This division of the State embraces the country lying between the heads of tide water in the East, and the Ocmulgee River in the West, and South to the corner of Liberty, Tattnall and Appling, while the Counties of Twiggs, Wilkinson, Washington, Glascock, Jefferson and Richmond, indicate its limits in the North. It differs from Middle Georgia in several important respects; its geological formations are tertiary instead of metamorphic: its average elevation is only about 250 feet above the sea; its surface is more level; its soils, for the most part, loamy or sandy; subsoil clay, red and yellow, four to six inches below the surface in clay lands, 8 to 12 inches in sandy lands; its forest growth is princially pine; it contains calcareous marls in considerable deposits. It is also the commencement of the section in which the sugar cane can be profitably cultivate l, while its rocks, which are few, are of a sedimentary character, with Iron ore and Buhr stone in several localities. Deposits of kaolin and pipe clay are found along its entire length from East to West. While pine is the leading forest growth, and the chief timber for building and export, there are also large tracts of oak and hickory. The soils in such localities are rather clayey or gray, chiefly the latter, and admirably adapted to the production of cotton and corn; cypress abounds in the swamps and lowlands. The County of Burke was, for many years, and until the late revolution in our system of labor, the leading cotton producing County of the State. The comparatively fresh lands of Decatur have, of late years, enabled her to claim and hold the championship in this particular product. Cotton, with corn, wheat (the adaptation to which lessens as we proceed southward into the pine lands,) oats, rye, barley, sugar cane, potatoes, constitute the staple products of the section.

The average yields per acre with fair culture, are: "Cotton, 350 lbs.; corn, 14 bushels; wheat, 12 bushels; oats, 25 bushels; cane syrup, 300 gallons; potatoes, 150 bushels; barley, 30 bushels. There is much high culture in the district, and these results are often quadrupled. The seasons for planting and harvesting are nearly the same as those of Middle Georgia, perhaps from 10 to 14 days earlier. The district is famous for its excellent fruit, especially peaches, strawberries and melons, large quantities of which are exported annually to Northern markets, Richmond, Burke and Washington being the principal counties engaged in the trade. The fig, grape—especially scuppernong—pear, plum, are all grown successfully. All the vegetables thrive well.

"The district is well watered, and water powers are ample for all purposes. The climate is perceptibly milder in Winter than that of Middle Georgia, and the average temperature in Summer higher; snows light, and only fall once in every four or five years. The average price of wood-lands in the oak and hickory section is $7 to $10 per acre, and improved lands $4 to $6; in the pine country uncleared lands can be bought from $1.25 to $2 per acre; improved farms from $5 to $10. Both can be had on liberal credit.

"The Bermuda and sedge grasses of the old fields in the upper tier of Counties, and the wire grass and cane of the Southern tier, afford the finest ranges for cattle and sheep the greater portion of the year. The Southern Counties abound in fish, deer and nearly every species. of wild game.

"This section of the State embraces the country lying between the heads of tide water in the East, and the Ocmulgee River in the West, and South to the corner of Liberty, Tatnall and Appling, and embraces the following Counties:

Bulloch,	Jefferson,	Pulaski,	Telfair,
Burke,	Johnson,	Richmond,	Twiggs,
Dodge,	Laurens,	Screven,	Washington,
Emanual,	Montgomery,	Tatnall,	Wilkinson.
Glasscock,			

SOUTHEAST GEORGIA.

This division embraces 15 counties, and comprises the coast and tide-water section of the State. The entire region is tertiary and mostly without rocks.

It has three distinct soils: 1, light, sandy; 2, dark sandy loam containing a large amount of vegetable matter; 3, reddish and clayey. The second variety is covered with a natural growth of yellow pine, magnolia, red bay, live-oak, cedar and cabbage palmetto, and in productiveness is excelled by no land in the State; it has a yellow clay subsoil, varying from 10 inches to 3 feet; Sea-island cotton, corn and sugar cane grow in the greatest luxuriance. The third variety is also very productive, pine, oak, hickory and gum being the prevailing forest growth; subsoil clay, red and yellow; average depth below the surface 8 to 12 inches. It is the great rice-producing section of the State—the broad bottoms of the Savannah, the two Ogeechees, the Altamaha and Satilla, being devoted almost exclusively to that cereal. It is also grown to a less extent on the St. Mary's, and considerable quantities on inland swamps, the irrigation in the latter being affected by means of "backwater," collected from rains and secured by dams. Sea-island, or long staple cotton, was the only variety formerly grown, but of late years the short staple has been introduced and cultivated with fair success. Corn, oats, pumpkins, potatoes, ground peas all do well. The sea-islands are devoted almost exclusively to cotton, corn, cane, fruits and vegetables. Cypress and palmetto abound in the swamps and river bottoms.

Average yield, per acre, of staple crops, with fair cultivation: Sea-island cotton, 600 lbs. in seed; corn, 15 bushels; oats, 25 bushels; rice, 40 bushels; cane syrup, 300 gallons; potatoes, 200 bushels. On best lands—1,500 lbs. seed cotton, 60 bushels rice, 600 gallons syrup, 50 bushels corn, 40 bushels oats, 400 bushels potatoes—are often produced on one acre. Corn planted middle of February till 1st of June, gathered in August and September; cotton planted March and April, gathered in autumn months; rice planted March to June, harvested last of August till 1st of October; cane planted February and March, cut in October and early in November; potatoes planted March to June, gathered July to November; oats sowed in October, harvested in May.

This section comprises the coast and tide-water section of the State, and embraces 15 counties, as follows:

Appling,	Chatham,	Effingham,	Pierce,
Bryan,	Clinch,	Glynn,	Ware,
Camden,	Coffee,	Liberty,	Wayne,
Charlton,	Echols,	McIntosh,	

SOUTHWEST GEORGIA.

This division is composed of 33 counties, and embraces all that country lying between the Ocmulgee and Allapaha Rivers in the East and the Chattahoochee River in the West; the Northern boundary being a line from Macon to Columbus, and the State of Florida its boundary in the South. Like Southeast Georgia the entire region is tertiary. It is more broken or rolling than Southeast Georgia, and with the exception of marl, buhr and limestone, is in a great measure destitute of rocks. It has also a greater proportion of clay lands and oak and hickory forest growth, although much the larger part of it is a light sandy soil, and was originally covered with yellow, or long leaf pine.

The clay lands are, generally, very rich, and their fertility lasting; the pine lands produce freely, are easily worked, but are less durable. The district contains very little waste land, or lands too poor or too swampy for cultivation, while the alluvial lands of the Chattahoochee and Flint Rivers, and of many of the creeks, have made the section famous as the best cotton regions of the State. Corn, oats, wheat, rye and sugar cane grow well.

The depth of the subsoil beneath the surface, on clay lands, is 6 to 10 inches; on sandy lands, from 12 inches to 3 feet. The preponderating forest growth is long leaf, or yellow pine, furnishing the best of lumber, large quantities of which are prepared annually for export and domestic use. The supply would seem to be almost inexhaustible.

Spirits of turpentine, rosin, pitch and tar—all the products of this tree—are made in considerable quantities, and the interest is on the increase. In the swamps and river bottoms there are cypress, cotton-wood, poplar, ash, maple, beach, birch, red-bay, magnolia, sweet-gum and water oak; while the growth of the clay belts is red and post oaks, black jack, hickory, walnut, black-gum, dogwood and buck-eye.

Cotton is the leading market crop of this division, and previous to the derangement of plantation labor by emancipation, its crop of the staple probably equalled the production of all the rest of the State. Corn and oats grow to great perfection. Sugar cane is a successful crop throughout the section; tobacco in considerable quantities, is grown in the Southern counties.

The average yield, on inferior land, per acre, with good cultivation, are: cotton, 500 lbs. in seed; corn, 10 bushels; oats, 15 bushels; syrup, 200 gallons; sweet potatoes, 150 bushels; ground peas, 50 bushels. On best lands, without manure, 1,500 to 2,000 lbs. cotton in seed, 50 to 75 bushels corn, 50 to 65 bushels oats, 400 gallons of syrup, and 400 bushels sweet potatoes, are often produced. It is reliably reported that a Berrien county farmer produced 800 bushels of sweet potatoes under high cultivation. Over 900 gallons of syrup, per acre, has been made in Thomas county.

This section embraces all that country lying between the Ocmulgee and Allapaha rivers in the East, and the Chattahoochee river in the West; the Northern boundary being a line from Macon to Columbus, and the State of Florida its boundary in the South. It comprises the following counties:

Baker,	Decatur,	Macon,	Stewart,
Berrien,	Dooly,	Marion,	Sumter,
Brooks,	Dougherty,	Miller,	Taylor,
Calhoun,	Early,	Mitchell,	Terrell,
Chattahoochee,	Houston,	Muscogee,	Thomas,
Clay,	Irwin,	Quitman,	Webster,
Colquitt,	Lee,	Randolph.	Wilcox,
Crawford,	Lowndes,	Schley,	Worth.

NOTE.—A proportion of the lands have suffered temporary exhaustion by in-judicious culture which claimed everything from the soil and returned nothing. This ruinous practice is fast giving way to a more enlightened and economical system. It has been ascertained that no soils on the continent are more sus-ceptible of recuperation and respond so bountifully to generous treatment. The abandoned fields, grown up in stunted pines, and for twenty or forty years considered useful only as pasturage, have been restored to cultivation, and are now among the most productive lands of the State.

Next to the manufacture of lumber, rice planting is the principal business of the coast counties of Georgia, giving employment to thou-sands of laborers who receive their wages weekly, and furnishing the merchants with a good local trade all the year. On the richest and most improved soils, fifty bushels of rice per acre is not an uncommon yield.

Cut a little green, it serves the place of hay and grain, for horses, mules, or cattle, and requires neither threshing nor hulling. Properly grown, it will produce on our ordinary pine lands, twenty to fifty bushels to the acre, which is worth in the hull at least $1 per bushel.

UPLAND RICE.

It has been a popular idea that rice could be grown only on lands, where it could be flooded at will. This, however, is a mistake, and it is a fact that any of our ordinary lands—pine lands or hammocks, wet land or dry land,—will, if properly planted and cultivated, grow a paying crop of upland rice. We have seen fair results on new pine land, but we have seen better crops on older lands.

The prices of all farm products are greater in Georgia than in the Western States.

For prices of land, water-powers, etc., etc., in any part of Georgia, in large or small tracts, address,

COMMISSIONER LAND & IMMIGRATION,
Atlanta, Georgia.

FARMING IN GEORGIA COMPARED TO FARMING IN OTHER STATES.

The writer prefers to use the testimony of Northern and foreign-born witnesses to prove that Georgia offers greater advantages than the West or North to the immigrant, rather than to indulge in any rounded periods. From the Superintendent at Castle Garden I received several publications in English and German. From one of these, printed in German, the following extract is taken:

"MOUNTAIN LAKE, Minnesota, 5 December, 1879.
"To MY MENNONITE BRETHREN IN RUSSIA:

" I have been living for four years in the State of Minnesota, United States of North America. The country is chiefly prairie, and one finds very little timber land. All wood required for building houses or fences is brought to my place of residence by railway, a distance of 135 to 200 miles. Hay is used as fuel. The winter is long and severe. Cattle must be fed for at least six months of the year. The field work is done chiefly by machinery, and the outlay of large sums of money is required in order to provide the necessary working materials, and work animals, which range at about following prices: breaking plough, $25 to $30; ordinary ploughs, $15 to $25; drill machines, $60 to $75; self-binding grain reaping machines, $285; mowing machine, $90; hay rake, $30; separators and threshing machines, $500 to $700; cleaning mill, $20 to $30; wagons, $65; oxen, $90 to $125; horses, $150 to $200 a pair. A great part of the summer is employed in providing winter food for the animals, and half of the winter is lost for field work by reason of unfavorable weather.
"ABRAHAM PENNER."

In Georgia, $12 will purchase the farm implements generally needed
by one laborer to cultivate the usual crops for one year.

COTTON PRESS IN THE COUNTRY.

The farmer may haul his cotton to the neighboring "cotton gin,"
and by paying a slight "toll," have his cotton ginned and baled for
market.   A two-horse wagon will haul easily three bales of cotton,
worth at present prices $165.00, at one load.   As the cost of handling
and hauling the grass and grain crops of the West is the greatest expense incidental to farming there, the above statement will be appreciated by the Western farmer.   The expense of transporting his
"money crop," to market is of prime importance to the farmer in
any latitude or country, and the cotton crop is certainly the most
economical field crop grown when viewed in this light.

COTTON PRESS.

The following from the *Atlantic Monthly,* the leading magazine of
Massachusetts, for January, 1880, is applicable to Minnesota, Dakota,
and other States and Territories of the Northwest :

"The imperative laws of the seasons have limited the time for the ef-
fective industry of the farmer to about one-fourth part of the year,
during which time the small farmer must make provision for all his force
for the full year, and from the fruit of the labor of himself and his own
family solely, during seed-time and harvest, must provide for all their

wants and comforts until the return of those seasons. * * * While the small farmer is compelled to feed, clothe, shelter, and altogether provide for the same number of persons for the whole year, the capitalist feeds, clothes and shelters only about one-fourth of the number in proportion to the amount of work done, and that for less than one-fourth of the year.   Against the unlimited use of this combination of capital, machinery and cheap labor, the individual farmer, either singly or in communities, cannot successfully contend, and must go under. * * * I particularly noticed the conspicuous absence of women and children on the large farms.   In no case was the permanent residence of a family to be found upon them, nor anything that could be called a home.   * * * While the capitalistic farmers are making colossal fortunes at seventy cents per bushel for wheat, the small farmers, depending mainly on their own labor, with limited capital and less machinery, are not making a comfortable subsistence, but are running behindhand, and must go under, and a further reduction in the market price of food products must hasten their end."

As a contrast, the following statement, taken from the *New York Sun*, February 1st, 1880, is offered:

"A farmer wrote to the Minneapolis (Minnesota) Convention that it cost him $9.10 to produce an acre of wheat, $41.55 for an acre of sorghum (amber cane), $8.95 for an acre of corn (maize).   But he expected to sell an acre of wheat for $15.00, an acre of sorghum for $75.00, an acre of corn for $10.00."

"Mr. John J. Parker, in Thomas County, (Southern Georgia,) produced in 1874, on one acre, 694½ gallons of cane syrup, at a cost of $77.50.   The syrup at 75 cents per gallon, the market price, brought $520.87; net profit from one acre of $443.37.   Southern Georgia is the 'home' of the sugar cane."

The former is probably an average crop, while the latter is exceptional.   The important question is, can like results be produced in Minnesota if the highest culture and treatment is given to products of the soil?   The best results yet attained there are probably found in the speech of Hon. Seth W. Kenney, who introduced the Early Amber cane in Minnesota, and who recently addressed a convention of farmers concerned in sugar production in that State.   I make the following interesting extract from his speech:

"The question is often asked me, if it will pay to raise sugar cane. I answer, yes.   The average product on our Minnesota land, with good, clean cultivation, is 160 gallons dense syrup per acre; that is, with good machinery to work it up, so that there is no unnecessary loss.   With a good granulating house, if properly defecated, 6 lbs. per gallon of good syrup can be obtained, which would be 960 lbs. for one acre, worth at least 8c. per lb., which would amount to $76.80; there would then be left ninety-three gallons of syrup, which at the low price of 30c. per gallon, would be $27.90, added to the sugar product would be $104.70, as the product of one acre."

Most of the "crops" of Georgia are cultivated with a single plow, the cost of which need not exceed three dollars.   As an additional evidence of what is done by Georgia farmers with one plow and two laborers, we append the following statements, which are not deemed extraordinary. S. W. McClendon, of Talbot County, made in the year 1879, twelve bales of cotton and three hundred bushels of corn, with one plow and a little hired labor. Sterling Jenkins, of same county, made fourteen bales to the plow. A. J. Jenkins, of Muscogee County, made twelve bales of cotton, two hundred bushels of corn, three hundred bushels of oats, two hundred bushels of potatoes, and cane enough to make several barrels of syrup, with one plow. C. D. McClendon, of Harris County, made with two plows thirty bales of cotton, one hundred and fifty bushels of corn, five hundred bushels of oats, one hundred bushels of potatoes.

According to the Sparta *Ishmaelite*, "Mr. Jas. H. Mitchell, of Hancock, made last year on a two mule farm, 28 bales of cotton, 500 bushels of corn, 300 bushels of potatoes, 174 gallons of syrup, and wheat, oats, peas, and other farm products in proportion.

(WESTERN GEORGIA.)

The *Columbus Enquirer* states that Z. H. Whittlesy made this year with one plow, ten bales of cotton weighing 500 pounds each, 600 bushels of corn, 200 bushels of potatoes, 300 bushels of oats, 3,000 pounds of fodder, 200 gallons of syrup, besides a good pea crop, and raising his meat. Cotton was worth on this farm 12 cents per pound; corn, 75 cents per bushel; oats, 60 cents per bushel; fodder, $1.25 per cwt.; syrup, 35 cents per gallon.

(SOUTHERN GEORGIA.)
(*From the Hawkinsville Dispatch*).

"Mr. Henry B. Marr, of this county, made last year upon his plantation, with one plow (or one mule) twenty-four bales of cotton, besides the usual small side crops of peas and potatoes."

A hundred such cases might easily be mentioned. Corn is priced in Georgia to-day at 75 cents per bushel, and is sold in some instances at $1.25 per bushel, "on time;" oats, at 65 cents; cotton, at 12 cents per pound, (there are 500 pounds in a bale of cotton); potatoes at 40 to 50 cents per bushel.

(NORTH GEORGIA.)
(*From the Marietta Journal*, 1880.)

"Five years ago, Mr. J. I. Chamberlain, of New York, came to Cobb County, and bought 500 acres of land from Mr. J. T. Burk-halter, on Powder Springs road, about five miles from Marietta. He came South in search of health. He first stopped at Knoxville, Tenn., and then tried Bartow County, but is now satisfied that the climate here suits him better, the water is purer, and the atmosphere more invigorating. He was embarrassed some when he took hold of the Burkhalter farm, but he went to work determined to succeed. The first year the land yielded poorly, and he only made sixteen bales of cotton. He began to plow deep, save manure and apply it to the land, and kept up a constant improvement. This last year he had 110 acres in cultivation and upon it made seventy-seven bales of cotton, averaging 492 pounds each. This was an astonishingly large yield, and the result was he cleared $3,000, He placed 16 tons of wheat bran on the land, at a cost of $14 per ton, and used 2½ tons of guano. He is out of debt, has spent $2,000 in improving his farm, has fine stock, good pastures, believes in the stock law, and in coming South has found both health and wealth. He has demonstrated the fact that there is as much in the man as in the land. This year he will use ten tons of guano, and we shall watch to see if he doesn't make over 100 bales and a profit of $4.000."

PRICE OF COUNTRY AND FARM PRODUCE IN GEORGIA AND IN THE NORTHWEST.

(From the Savannah, Ga., Morning News, March 12, 1880).

Country Produce.

Grown Fowls, ℔ pair	$0 50	@$0 60
Half-grown, ℔ pair	30	@ 35
Three-quarters grown, ℔ pair	40	@ 45
Ducks, Muscovy, ℔ pair	85	@ 1 00
Ducks, English, ℔ pair	55	@ 65
Turkeys, ℔ pair	1 50	@ 2 50
Chickens, dressed, ℔ lb	11	@ 12½
Turkeys, dressed, ℔ lb	12½	@ 15
Eggs, country, ℔ doz	15	@ —
Eggs, Western, ℔ doz	15	@ —
Butter, country, ℔ lb	15	@ 25
Peanuts, Georgia, ℔ bushel	1 15	@ 1 25
Peanuts, Tennessee, ℔ bushel	1 10	@ 1 20
Florida Sugar, ℔ lb	5	@ 6¼
Florida Syrup, ℔ gallon	40	@ 45
Honey, ℔ gallon	45	@ 65
Irish Potatoes, ℔ barrel	2 25	@ 2 50
Sweet Potatoes, ℔ bushel	40	@ 50

RICE.—We quote rough rice:

Prime Lots (tide water)	1 40	@ 1 60
Country Lots	1 25	@ 1 35

HIDES, ETC.—Hides—Receipts small and prices declined. We quote: Dry flint, 16c.; salted, 12@14c. Tallow, 6c.; wax, 22c.; deer skins, 40c.; otter skins, 25c.@$2.00.

HAY.—Northern, $1.05@$1.10 wholesale; Eastern and Pennsylvania, $1.15@$1.30.

WESTERN MARKETS.

(From the St. Paul (Minnesota) Pioneer Press, March 12, 1880.)

Country Produce.

BUTTER.—Choice grades are selling at 20@22c.; ordinary roll at 16@18c., and common at 12@14c. Grease, 8@10c.

EGGS—Firm at 11c. per dozen.

POTATOES—In demand at 25@30c. for early rose.

ONIONS—$4.50@$6.00 per barrel, for good to choice.

APPLES—$4@$5 per barrel.

HONEY—Sells slowly at 14@15c. for choice.

BEANS—In good supply and dull at $1@$1.40 for poor to common; hand picked medium, $1.60@$1.70; hand picked navy, $1.80.

MOLASSES AND SYRUP.—Common molasses, sells at 30c.; New Orleans, do., fair, 50c.; New Orleans, do., choice, 60c.; syrup, fair, 40c.; syrup, good, 50c.; 3c. additional on half barrels; 5c. additional on five and ten gallon kegs.

CHEESE.—Good cream, 14@15c.; half skim, 11@12c.

Live Stock.

CATTLE. -There was an active demand for good stock. Two car loads of Minnesota steers were received from St. Peter. One load was sold at $4.25 per cwt.

HAY—Quiet and unchanged; 70@75c. ℔ cwt.

HIDES AND FURS—Market quiet and steady; green butchers' hides, 7@7½c.; green salted, 8@8½c.; calf, 10@12c.; dry salt, 10@12c.; flint, 12@14c.; furs steady, unchanged; casses would bring 7½c.

DRESSED POULTRY—Turkeys are in active demand at 11@11½c.; chickens in fair request at 10@10½c. Ducks and geese are quotable at 6@8c.

CORN—Very quiet; only moderate inquiry; held at 34@35c. for car lots on track.

OATS—Steady and firmly held at 31@32c. for No. 2, and 32@34c. for choice white on track.

RYE—Steady sales; car lots on track at 50@55c.; lots by teams, 45@50c.

The following table shows the price of farm products at Chicago, St. Paul and Savannah, March 12, 1880:

	Chicago.	St. Paul.	Savannah.
Corn	37¼c	34c.	77c.
Oats	—	47@48c.	57@60c.
Hay, ℔ cwt	—	70@75c.	$1.05@$1.10

Table showing the Population of each County and County Site by the Census of 1870, and the Average Value of Improved and Wild Lands, according to the Comptroller-General's Report for 1876; also the Railroad traversing each County.

COUNTIES.	POPULATION OF EACH COUNTY.			County Sites.	Population of County Sites.	Average value of improved lands.	Average value of wild lands.	RAILROAD TRAVERSING EACH COUNTY.	Proportion of white laborers engaged in farm work.
	White.	Colored.	Aggregate.						
Appling.........	4,110	976	5,086	Holmesville........	*	$0.59	$0.15	Macon and Brunswick.	75 per cent.
Baker.........	1,888	4,955	6,843	Newton........	145	1.93	.27	Atlantic and Gulf.	25 "
Baldwin.........	3,844	6,774	10,618	Milledgeville........	2,750	3.43	.10	Macon and Augusta; also a branch of the Central.	10 "
Banks.........	4,052	921	4,973	Homer........	120	3.55	.72	Atlanta and Richmond Air-line.	75 "
Bartow.........	11,846	4,719	16,565	Cartersville........	2,232	7.30	.46	Western and Atlantic; also the Cherokee Railroad.	75 "
Berrien.........	4,057	460	4,517	Nashville........	95	1.10	.42	Albany and Brunswick.	87 "
Bibb.........	9,831	11,424	21,255	Macon........	10,810	9.86	.19	Central: also Macon and Augusta, Macon and Brunswick, and Southwestern.	10 "
Brooks.........	4,111	4,231	8,342	Quitman........	784	3.00	.32	Atlantic and Gulf.	38 "
Bryan.........	1,647	3,605	5,252	Eden........	*	1.31	.60	" "	55 "
Bullock.........	3,866	1,744	5,610	Statesborough...	33	.88	.36		44 "
Burke.........	4,243	13,436	17,679	Waynesborough...	843	2.95	.54	Central Railroad.	24 "
Butts.........	3,496	3,445	6,941	Jackson........	*	4.16	.54		50 "
Calhoun.........	2,026	3,477	5,503	Morgan........	126	2.23	.41	Albany and Blakely Extension of Southwestern Railroad.	25 "
Camden.........	1,458	3,157	4,615	St. Mary's........	702	1.77	.39		No returns made.

* Population not given separately.

Table showing the Population of each County, etc.—(*Continued.*)

COUNTIES.	POPULATION OF EACH COUNTY.			County Sites.	Population of County Sites.	Average value of improved lands.	Average value of wild lands.	RAILROAD TRAVERSING EACH COUNTY.	Proportion of white laborers engaged in farm work.
	White.	Colored.	Aggregate.						
Campbell.............	6,589	2,587	9,176	Fairburn............	305	$6.01	$0.78	Atlanta and West Point.	No returns made.
Carroll...............	10,473	1,309	11,782	Carrollton..........	‡	4.66	.58	Griffin and North Alabama.	80 per cent.
Catoosa	3,793	616	4,409	Ringgold.......	316	6.11	1.12	Western and Atlantic, or State Road.	85 ..
Charlton	1,496	401	1,897	Trader's Hill......	‡	.60	.06		90 ..
Chatham.............	16,760	24,518	41,278	Savannah...........	28,235	9.12	.11	Central Railroad; also Atlantic and Gulf, and the Charleston and Savannah.	10 ..
Chattahoochee......	2,654	3,405	6,059	Cusseta............	216	2.63	.25		44 ..
Chattooga..........	5,399	1,503	6,902	Summerville	281	4.66	.44		80 ..
Cherokee............	9,117	1,281	10,399	Canton	214	4.14	.55		78 ..
Clarke*...............	6,488	6,453	12,941	Athens	4,251	6.19	.20	Athens branch of Georgia Railroad; also Northeastern Railroad.	30 ..
Clay	2,644	2,849	5,493	Fort Gaines........	758	2.60	.57	Fort Gaines branch of Southwestern Railroad.	44
Clayton.............	3,734	1,743	5,477	Jonesborough.......	531	7.55	.59	Macon and Western (branch of Central).	75 ..
Clinch..............	3,437	517	3,945	Homerville.........	‡	.90	.15	Atlantic and Gulf; also Florida connection of same.	75 ..
Cobb................	10,593	3,217	13,810	Marietta............	1,888	7.62	.55	State Road.	60 ..
Coffee..............	2,514	678	3,192	Douglas............	‡	.50	.23	Albany and Brunswick.	20 ..
Columbia†..........	4,080	9,449	13,529	Appling............	‡	3.66	.31	Georgia Railroad.	No returns made.
Colquitt	1,517	137	1,654	Moultrie............	‡	.84	.21		96 per cent.

* Clarke also included Oconee County in 1870.   † Columbia also included McDuffie County in 1870.   ‡ Population not given separately.

County	White	Colored	Total	County Seat	Pop.			Railroads	Per cent
Coweta	7,856	8,019	15,875	Newnan	1,917	.38	4.49	Atlanta and West Point.	31 per cent.
Crawford	3,284	4,273	7,557	Knoxville	223	.34	2.58	Southwestern.	40 "
Dade	2,788	245	3,033	Trenton	§	.33	6.26	Alabama and Chattanooga.	80 "
Dawson	4,032	337	4,369	Dawsonville	§	.07	3.04		90 "
Decatur	7,465	7,718	15,183	Bainbridge	1,351	.22	1.28	Atlantic and Gulf.	25 "
De Kalb	7,352	2,662	10,014	Decatur	401	.70	9.19	Georgia Railroad: also Atlanta and Richmond Air-line.	75 "
Dodge*				Eastman	§	.47	1.25	Macon and Brunswick.	
Dooly	4,935	4,855	9,790	Vienna	§	.19	2.64		50 "
Dougherty	2,093	9,424	11,517	Albany	2,101	.33	3.23	South Georgia and Florida (branch of the Atlantic and Gulf).	30 "
Douglas†				Douglasville	§	.26	4.18		5 "
Early	2,826	4,172	6,998	Blakely	§	.32	1.62	Albany and Blakely extension of South-western Railroad.	60 "
Echols	1,513	465	1,978	Statenville	61	52	1.01	Florida connection of Atlantic and Gulf.	25 "
Effingham	2,507	1,704	4,211	Springfield	§	.32	1.29	Central Railroad.	75 "
Elbert	4,386	4,863	9,249	Elberton	§	.33	3.86	Elberton Air-line Railroad.‖	57 "
Emanuel	4,431	1,703	6,134	Swainsborough	108	.53	.94		36 "
Fannin	5,285	144	5,429	Morganton	§	.17	1.67		82 "
Fayette	5,683	2,538	8,221	Fayetteville	§	1.01	2.07	Augusta and Knoxville. Not finished.	99½ "
Floyd	11,473	5,757	17,230	Rome	2,748	.50	6.91	Griffin and North Alabama. Selma, Rome, and Dalton; Rome and Kingston; Cherokee Railroad.	60 "
Forsyth	6,862	1,121	7,983	Cumming	267	.45	4.59		53 "
Franklin	6,034	1,859	7,893	Carnsville	266	.63	3.91	Elberton Air-line,‖	90 "
Fulton	18,164	15,282	33,446	Atlanta‡	21,789	.14	14.29	Georgia Railroad; State Road; Macon and Western; Atlanta and West Point; Atlanta and Richmond Air-line.	75 "
Gilmer	6,527	117	6,664	Ellijay	§	.29	1.47		55 "
Glascock	1,917	819	2,736	Gibson	§	.15	3.10		No returns made.
Glynn	1,926	3,450	5,376	Brunswick	2,343	.61	3.42	Albany and Brunswick; Macon and Brunswick.	98 per cent.
Greene	4,293	8,156	12,454	Greensborough	913	.30	4.45	Georgia Railroad; also Athens branch.	10 "
Gordon	7,726	1,536	9,268	Calhoun	427	.50	6.33	State, or Western and Atlantic; also Selma, Rome, and Dalton.	70 "

* Cut off from Telfair, Pulaski, and Montgomery since 1870.   † Cut off from Campbell and Carroll since 1870.   ‖ Now being built.
‡ According to census taken in 1877, 35,956.   § Population not given separately.

Table showing the Population of each County, etc.—(Continued.)

Counties.	Population of each County. White.	Colored.	Aggregate.	County Sites.	Population of County Sites.	Average value of improved lands.	Average value of wild lands.	Railroad traversing each County.	Proportion of white laborers engaged in farm work.
Gwinnett	10,272	2,159	12,431	Lawrenceville	*	$4.46	$0.59	Atlanta and Richmond Air-line. also	75 per cent.
Habersham	5,373	949	6,322	Clarkesville	263	1.90	.36	Atlanta and Richmond Air-line; also Elberton Air-line.	90 "
Hall	8,317	1,290	9,607	Gainesville	472	3.56	.43	Atlanta and Richmond Air-line; also Northeastern.	90 :
Hancock	3,645	7,672	11,317	Sparta	*	4.24	.40	Macon and Augusta.	20 "
Haralson	3,685	319	4,004	Buchanan	*	2.92	.36		92½ "
Harris	5,791	7,493	13,284	Hamilton	359	4.32	.39	North and South Georgia Railroad.	34 "
Hart	4,841	1,942	6,783	Hartwell	154	3.47	.30	Elberton Air-line.	59 "
Heard	5,218	2,648	7,866	Franklin	*	3.66	.72		62 "
Henry	6,269	3,833	10,102	McDonough	320	5.10	.69	Macon and Western.	50 "
Houston	5,071	15,332	20,406	Perry	836	4.46	.31	Southwestern.	25 "
Irwin	1,541	296	1,837	Irwinville	*	.90	.16	Albany and Brunswick.	75 "
Jackson	7,471	3,710	11,181	Jefferson	:	4.03	.32	Northeastern.	66 "
Jasper	3,384	6,555	10,439	Monticello	*	2.89	.35		33 "
Jefferson	4,247	7,943	12,190	Louisville	356	3.07	.35	Central Railroad.	26 "
Johnson	2,049	915	2,964	Wrightsville	*	1.57	.52		80 "
Jones	2,991	6,445	9,436	Clinton	362	3.56	.33	Central Railroad; also Macon and Augusta.	16 "
Laurens	4,180	3,654	7,834	Dublin	*	1.35	.36		42 "
Lee	1,924	7,643	9,567	Starkville	*	2.99	.47	Southwestern Railroad.	5 "
Liberty	2,428	5,260	7,688	Hinesville	*	.80	.50	Atlantic and Gulf.	25 "
Lincoln	1,797	3,616	5,413	Lincolnton	92	2.85	.36		36 "

* Population not given separately.

County	White	Col'd	Total	County Seat	Pop.			Railroad	Per cent
Lowndes	4,276	4,045	8,321	Valdosta	1,199	1.77	.13	Atlantic and Gulf.	10 per cent.
Lumpkin	4,699	462	5,161	Dahlonega	471	1.71	.29	Southwestern.	90 "
Macon	3,975	7,843	11,458	Oglethorpe	400	2.87	.36		25 "
Madison	3,646	1,581	5,227	Danielsville	§	3.27	.58		65 "
Marion	4,169	3,830	7,999	Buena Vista	525	2.55	.23	Georgia Railroad.	22
McDuffie*	1,196	3,288	4,491	Thomson	369	3.64	.26	Atlantic and Gulf.	No returns made.
McIntosh				Darien	547	2.94	.39		5 per cent.
Meriwether	6,387	7,369	13,756	Greenville	§	3.12	.24		32 "
Miller	2,135	956	3,091	Colquitt	126	1.25	.48	South Georgia and Florida.	25 "
Milton	3,818	466	4,284	Alpharetta	289	5.45	.76	Macon and Western.	75 "
Mitchell	3,683	2,950	6,633	Camilla	§	2.49	.55		25 "
Monroe		10,804	17,213	Forsyth	1,389	4.50	.36		20 "
Montgomery	2,478	1,108	3,386	Mount Vernon	248	.65	.11	Georgia Railroad.	72 "
Morgan	3,637	7,058	10,696	Madison		4.45	.31		24 "
Murray	5,743	757	6,500	Spring Place		4.67	.69	Southwestern; also North and South Georgia.	80 "
Muscogee	7,441	9,220	16,663	Columbus	7,401	6.40	.52	Georgia Railroad.	15 "
Newton	8,601	6,014	14,615	Covington	1,121	5.42	.36		45 "
Oconee†				Watkinsville	643	4.42	.29		No returns made.
Oglethorpe	4,641	7,141	11,782	Lexington		3.58	.27	Athens branch of Georgia Railroad.	43 "
Paulding	7,083	556	7,639	Dallas	§	5.03	.32		90 per cent.
Pickens	5,188	129	5,317	Jasper	§	2.40	.62		83 "
Pierce	1,964	814	2,778	Blackshear	490	.62	.24	Atlantic and Gulf.‖	52 "
Pike	5,999	4,906	10,905	Zebulon		5.42	.44	Macon and Western; also Upson County Railroad.	
Polk	5,244	2,578	7,822	Cedartown	323	6.73	.45	Cherokee Railroad.	57½ "
Pulaski	5,955	2,985	11,940	Hawkinsville	813	1.81	.44	Macon and Brunswick; also Hawkinsville branch.	27 "
Putnam	3,016	7,445	10,461	Eatonton	1,240	4.22	.36	Milledgeville and Eatonton.	18 "
Quitman	1,773	2,377	4,150	Georgetown	263	2.73	.18	Branch of the Southwestern.	22 "
Rabun	3,137	119	3,256	Clayton	70	.94	.11	Northeastern.‖	92 "
Randolph	5,084	5,477	10,561	Cuthbert	2,210	2.84	.36	Branch of the Southwestern.	25 "
Richmond	13,157	12,565	25,722	Augusta‡	15,389	8.92	.35	Georgia, Central, Port Royal, South Carolina, Charlotte, Columbia, and Augusta, and Augusta and Knoxville, now being built.	65 "

* Cut off from Columbia since 1870.   † Cut off from Clarke since 1870.
‡ According to census taken in 1877, 23,768.   § Population not given separately.
‖ Now being extended to Rabun.

Table showing the Population of each County, etc.—(Continued.)

Counties.	Population of each County.			County Sites.	Population of County Sites.	Average value of improved lands.	Average value of wild lands.	Railroad traversing each County.	Proportion of white laborers engaged in farm work.
	White.	Colored.	Aggregate.						
Rockdale*				Conyers	637	$7.25	$0.57	Georgia Railroad.	37 per cent.
Schley	2,278	2,851	5,129	Ellaville	157	3.34	.32	Central Railroad.	37 "
Scriven	4,287	4,888	9,175	Sylvania	‡	1.07	.33		44 "
Spalding	5,327	4,878	10,205	Griffin	3,421	5.77	.14	Macon and Western.	43 "
Stewart	5,104	9,100	14,204	Lumpkin	778	3.62	.47		30 "
Sumter	5,920	10,639	16,559	Americus	3,259	3.77	.39	Southwestern.	18 "
Talbot	4,761	7,152	11,913	Talbotton	796	3.34	.09	Southwestern.	No returns made.
Taliaferro	1,809	2,987	4,796	Crawfordville	‡	3.10	.24	Georgia Railroad; also Washington branch.	70 per cent.
Tattnall	3,580	1,280	4,860	Reidsville	55	.72	.33		68 "
Taylor	4,181	2,962	7,143‡	Butler	‡	2.13	.35	Southwestern.	50 "
Telfair	2,100	1,145	3,245	McRae	‡	1.01	.27	Macon and Brunswick.	50 "
Terrell	3,769	5,284	9,053	Dawson	1,099	2.61	.32	Southwestern.	25 "
Thomas	6,160	8,363	14,523	Thomasville	1,651	2.49	.33	Atlantic and Gulf; also South Georgia and Florida.	42 "
Towns	2,623	155	2,780	Hiawassee	‡	1.95	.17		98 "
Troup	6,408	11,224	17,632	La Grange	2,053	4.52	.25	Atlanta and West Point; also North and South Georgia.	19 "
Twiggs	2,913	5,632	8,545	Marion	265	2.11	.29		30 "
Union*	5,153	114	5,267	Blairsville	‡	1.56	.16	Macon and Brunswick.	95 "
Upson	4,865	4,565	9,430	Thomaston	630	3.80	.51	Upson County Railroad.	43 "

* Cut off from Newton since 1870.
‡ Population not given separately.

† There is also in this county the town of West Point, with a population of 1405 in 1870.

County				Town				Railroad	per cent.
Walker	8,396	1,529	9,925	Lafayette	251	5.17	.46	Selma, Rome, and Dalton.	87 per cent.
Walton	6,876	4,162	11,038	Monroe	438	4.99	.22	Georgia Railroad.	47 "
Ware	1,834	452	2,286	Waresborough	*	.70	.09	Atlantic and Gulf.	75 "
Warren	4,285	6,260	10,545	Warrenton	630	3.91	.41	Georgia R.R.; also Macon and Augusta.	34 "
Washington	7,530	8,312	15,842	Sandersville	*	3.70	.33	Central Railroad.	19 "
Wayne	1,798	379	2,177	Waynesville	*	.47	.31	Atlantic and Gulf, Macon and Brunswick, Albany and Brunswick.	75 "
Webster	2,439	2,238	4,677	Preston	186	3.03	.53		42 "
White	4,042	564	4,606	Cleveland	145	2.31	.17		99 "
Whitfield	8,606	1,511	10,117	Dalton	1,809	5.82	.37	Western and Atlantic or State; Selma, Rome, and Dalton; East Tennessee and Georgia	82½ "
Wilcox	1,902	537	2,439	Abbeville	*	.95	.35	Washington branch of Georgia Railroad.	50 "
Wilkes	3,069	7,827	11,796	Washington	1,506	4.24	.34		78 "
Wilkinson	4,684	4,699	9,383	Irwinton	241	2.49	.33	Central Railroad.	42½ "
Worth	2,673	1,105	3,778	Isabella	54	1.36	.21	Albany and Brunswick.	58 "
	638,926	545,142	1,184,109						

* Population not given separately

There are in Georgia about 2400 miles of railroad, of which the most extensive roads are the Georgia Railroad and branches, embracing 228 miles; the Central Railroad and branches, with 388 miles; the Southwestern and branches, including 306 miles; and the Savannah, Florida and Western, with its branches, 345 miles.

UPLANDS HOTEL, EASTMAN, DODGE COUNTY GEORGIA,

A TYPICAL GEORGIA VILLAGE.

(SOUTHERN GEORGIA.)

A forest in 1870; a village of about 800 inhabitants in 1880.

The climate of Southern Georgia has been fitly described by the following extract from :

"'Winter Homes for Invalids,' by Doctor Joseph W. Howe, professor of Clinic Surgery in the University of New York, in reference to the 'pine forests' of Southern Georgia.

"Pine grove localities have the reputation of being very healthy. There is usually complete freedom from malaria and pulmonary diseases. The atmosphere, impregnated as it is with the peculiar volatile principle of trees, has a soothing effect on inflamed throats and irritable lungs. The air agrees with everybody. Invalids with troublesome cough and shortness of breath rapidly improve after a short residence, and some far advanced in tubercular diseases recover their health completely. The dryness and mildness of the atmosphere has, of course, something to do with the beneficial effects experienced, but there is no doubt whatever that much of the benefit arises from the air being impregnated with the piney odor from the Ocean and Gulf."

Eastman, Dodge County, Georgia, followed the building of the Macon and Brunswick Railroad, through this attractive region, which has opened a market for the pine timber. Eastman is the county town, and for ten miles East and West, up and down the railroad, and for fifteen miles North to the Oconee River, and fifteen miles South to the Ocmulgee River, lies a region of beautifully undulating or rolling prairie land, well watered with springs and small clear streams, embracing several hundred thousand acres, all of which is commercially tributary to this rising town.

In a sanitary point of view, Eastman equals Aiken, S. C., or the Sand Hills near Augusta, Georgia, and is probably unsurpassed for healthfulness by any town in the United States. If we may judge of the immunity of any given locality from certain diseases, by the absence of those diseases, the region in which Eastman, Georgia, is located, is far superior as a sanitarium for persons suffering from catarrh, bronchial, nasal and aural, and tubercular diseases, to the elevated land of Colorado, Minnesota, Nebraska, and New Mexico. While these diseases are almost unknown to the natives of the pine uplands of Southern Georgia, the reports of eminent physicians show that they are very prevalent in the regions above alluded to.

RESIDENCE OF WM. PITT EASTMAN, ESQ.

The equability of temperature, and the inhalation of the pine odor at Eastman, Thomasville, and other winter resorts in Southern Georgia, and the absence of the sudden and great vicissitudes in temperature peculiar to great altitudes, makes the pine uplands of that region the most desirable in the Union for persons threatened with phthisis pulmonalis.

The air of these pine uplands is strongly recommended by Dr. Willard Parker of New York, and other physicians of the highest standing.

Eastman is situated about 500 feet above the level of the sea, on ground marked by inequalities of surface sufficient to provide ample drainage without being either hilly or level. The water is pure and good. The grounds about the hotel and court house form a fitting center to the town. Avenues to the number of eight are laid out at right angles from the railroad, while parallel with the railroad are other streets, named from the native forest trees.

The Court House was presented by Hon. Wm. E. Dodge of New York, for whom the county is named.

What has been accomplished at Eastman can be accomplished at many other points in Georgia. Population brings with it more or less capital, and capital cannot seek a more promising or safer investment than the selection and settlement of eligible sites in Georgia. Many desirable offers are made to this end.

A MOUNTAIN SCENE.

TYPICAL VALLEYS, (NORTHWEST GEORGIA.)

GRASS CULTURE.

Cedar Valley, Vann's Valley, and Texas Valley, in Northwestern Georgia, recall the most delightful rural landscapes in Pennsylvania, while the valley of the Etowah resembles the famed Mohawk Valley in New York. The mountain ranges of Northwestern Georgia gradually recede into foot-hills from the summit of any one of which may be seen, any summer day, fields of waving grain, and meadows of clover and grasses that rival any in the North. Throughout this productive region, numerous clear, rapid streams course from mountain to river. No part of America is better watered or more admirably suited for mixed husbandry and stock raising. Mosquitoes and troublesome insects are almost unknown, and the nights are delightfully cool. For eight months in the year the days are as perfect as can be found anywhere. The three valleys described are selected because they are typical of Northwestern Georgia, each being distant but a few miles from the other. They are also alluded to in Derry's "Georgia."

VANN'S VALLEY.

Vann's Valley begins at Rome—the leading town of this part of Georgia—and, after extending eighteen miles, abruptly terminates in a group of hills. Like most of the hills in this region, these are mainly left as nature made them, and the colors of the sky, transparent atmosphere, and the richly tinted verdure of these hard woods, combined with the ripening grain in the valley, form a beautiful landscape. This valley is from one to three miles in width, and is well improved, having many substantial farmers whose large barns demonstrate the productiveness of the dark red limestone soil. The lovely village of Cave Spring is in this valley, and the cave and spring from which it takes its name are also typical of this region. The spring gushes from the cave at the base of one of the hills that surround the village, in a stream so bold that in England it would be called a river. Few villages possess more picturesque surroundings, and it is here that the State has erected large and comfortable buildings as an Asylum for the deaf and dumb. Deprived of the two most priceless senses, that of sight is trebly valued and valuable, and here nature has lavishly bestowed its wealth, in brilliant flora, balmy sun-lit air, clear streams that flow over pebbly bottom, and a varied landscape of hill and dale, field and forest, sky and water that will leave indelible impressions upon each one who finds a pleasant refuge and congenial employment here. The grass lands in these valleys equal the famous blue-grass lands of Kentucky and Missouri.

CEDAR VALLEY.

Two miles distant, separated by hills whose timbered sides are full of iron ore, or marble, is Cedar Valley, long famous for its productive soil. Being much larger than Vann's Valley, the farms are larger, and extensive cotton fields, interspersed with meadows and grain fields, greet the eye. On the banks of Cedar Creek several extensive iron furnaces, for the manufacture of pig metal have been erected, while charcoal iron, owing to the proximity of lime-stone and abundant forest, is manufactured very profitably. The coal fields are not distant, and the supply of the necessary raw materials is thus practically unlimited. The Cherokee Railroad extends from Cartersville to Cedartown, a distance of forty miles.

TEXAS VALLEY.

Twelve miles from Rome, resembles the two named, with the difference that the soil is a gray loam, with much more sand than that in Vann's or Cedar Valley. These valleys are peopled with good society, who appreciate and have churches and public schools and are very hospitable. The valleys of Armuchee and Chattooga are equally attractive, while in Bartow, Gordon, Catoosa, and other counties of this part of Georgia, there are valleys that equal or excel those mentioned.

The valleys of Northern Georgia and the temperature and rain-falls are eminently adapted to grass culture. While three tons of clover or Timothy hay per acre, is an average annual crop in Upper Georgia, the following exceptional crops are reported and can be verified:

Mr. J. R. Winters, of Cobb county, (Upper Georgia) produced, in 1873, from 1.15 acres, 6,575 pounds of dry clover hay at the first cutting of the second year's crop.

Mr. T. H. Moore, of same county, produced on one acre 105 bushels of corn, while Mr. Jeremiah Daniel produced 125 bushels.

Mr. R. Peters, Jr., of Gordon county, (Upper Georgia) harvested, in 1874, from three acres of lucerne, four years old, fourteen tons and 200 pounds of hay, or 9,400 pounds per acre.

Mr. L. B. Willis, in Greene county, (Middle Georgia,) in June, 1873, from one acre and a third, harvested twenty bushels of wheat, and the following October, 27,130 pounds of corn forage. From the forage alone he received a profit of $159.22 per acre.

Dr. W. Moody, of the same county, harvested, at one cutting, from one acre of river bottom, in 1874, 13,953 pounds of Bermuda grass hay; cost, $12.87, value of hay, $209.29, net profit, $196.42.

Nine-tenths of the farmers in Georgia have been engaged in *killing grass* in order to raise cotton, but the hay crop is the most valuable one, time, money and labor considered, that can be raised in Georgia. That it pays there much better than in the North or West, the following market quotations will demonstrate:

St. Paul and Minneapolis Pioneer Press, Thursday, March 11, 1880.

HAY—Was dull at $7 per ton for baled wild on track, and $12 for tame.

(Courier Journal, (Louisville Ky.,) Wednesday Morning,
March 17.)

HAY—Steady at last prices. We quote prime to choice timothy at
$16@$18; mixed at $13@$16.

(The Morning News (Savannah, Ga.,) Friday, March 12, 1880.)

HAY—Demand good. We quote: Northern, $1.05@$1.10, wholesale;
Eastern and Pennsylvania, $1.15@1.30.

This difference of $9 per ton in favor of Georgia will be readily
appreciated by the practical farmer.

The dairy farmer who settles in Georgia will find a wide field for
profit, there being but little competition. The manufacture of cheese
is an unknown industry in Georgia.

A Scotch gentleman, representing a wealthy company in Scotland,
who proposes to purchase land and colonize it with Scotch working
people, writes as follows :

" You recommend purchasing a tract of valley land amongst higher
land of North Georgia, and state that good 'range' can be obtained
without charge. I prefer Northern to Southern Georgia, but it must
require some housing as well as feeding in sheds in Northern Georgia
none of which I take it, is required in Southern Georgia. Does not
the rainfall interfere with the hay harvest? If only five weeks feeding
of hay is required, *and hay can be bought at* $1.25 *per ton on Ranche,*
I prefer Northern Georgia, even though it is a few miles distant from
railways."

This letter is copied in order to illustrate the fact that hay is the
most valuable agricultural product, perhaps, in Georgia. The writer
has frequently sold large quantities of hay on his farm, not at $1.25
per ton, but at $1.25 *per* 100 *lbs. or* $25 *per ton.*

The clean culture necessary in cotton culture is an admirable preparation for grass growing, having the same effect as the "fallow,"
which is so potent a factor in English farming. We believe the price,
($1.25 per ton,) alluded to above, is the price on the Western prairies,
and we are at a loss when we attempt to reconcile this fact with the
alleged "high wages" paid the laborer in the prairie States. We do
not believe that a laborer in Georgia can be hired to cut, save and
deliver a ton of hay at $1.25 per ton.

The following from the *New York Bulletin,* will show that during
harvest time rainfall is not peculiar to Georgia :

(From our Crop Correspondent.)

" Chicago, July 9, 1880.

"Wet weather continues in Illinois ; so much so that all through
the corn belt, cultivation of this crop has been suspended for two
weeks. The weather has been so bad that it has been almost impossible to secure the hay crop in good condition, and farmers are very
much discouraged at the condition of things.

SHEEP HUSBANDRY IN GEORGIA.

CROSSING THE FORD.

The present high price of wool indicates scarcity, and a recent issue of the *Economist* suggests reflection. It is there stated that we consume 360,000,000 lbs. of wool per annum while we grow only 225,000,000 lbs. The price of wool in Georgia is forty cents a pound. The prime requisites for sheep husbandry, as a specialty, are fresh water, shade, and an abundance of pasture-land, that the needed change be given. In no part of the United States can extensive sheep-walks, possessing these requisites in equal extent, be purchased so cheaply as in Georgia, and in no State are the climatic and transportation advantages excelled. The habits of sheep in feeding early in the morning and late in the afternoon, and resting during the heated hours in Summer, demonstrate the necessity for shade. There are in Georgia about 10,000,000 acres of unemployed land, having ample shade. These lands are admirably adapted to sheep husbandry, a large part of them being worth more for timber purposes than the land will cost. Forty thousand acres may be bougnt for two dollars per acre, all of which is situated at an average distance of not more than ten miles from any navigable stream or railroad, and is only about forty hours distant from New York. Millions of acres of virgin soil, covered with grass the year round, with yellow pine forests and no undergrowth, presenting a park-like appearance to the eye, await the advent of the shepherd and farmer.   Streams abound there ; no shelter is needed in winter, and sheep, as a rule, cost nothing, except the salt they eat, the care they receive, and the cost of shearing. Georgia unwashed wool is said to be as clean as Pennsylvania brook-washed.   This is owing to the fact that it is free from hay seed—the wire grass being perennial—and that the heavy spring rains wash out the yolk and dirt just before shearing time.   Southern Georgia offers perhaps the best field for investment to wool growers on the continent. The climate of Georgia corresponds with that of the best wool growing countries.   The wire-grass and perpetual plant-growth in Southern Georgia, the Bermuda grass sod in Middle Georgia, and the sod of the English grasses, the clovers and blue grass in North Georgia, added to the Southern field pea, are alike well adapted to sheep husbandry.   Land that is not considered profitable for cultivation will support five sheep to the acre.   Very few farmers do more for their sheep than to mark, shear and salt them.

PROFITS OF SHEEP HUSBANDRY. .

The average annual cost per head of keeping sheep in Georgia, is only 54 cents.   The average cost of raising a pound of wool is only 6 cents, while the average price for which the unwashed wool is sold, is 33⅓ cents, or 27½ cents net.   The average yield of unwashed wool to the sheep is 3-44 pounds, which at 27½ cents net, gives an average clear income in wool from such sheep of 94 cents.   The average price for lambs sold to the butcher in Georgia is $1.87.   The average price of stock sheep is $2.68 per head.   The average price of mutton is $2.75 per head.   The average annual profit invested in sheep in Georgia is 63 per cent.   The following reports have been sworn to in the presence of disinterested parties.   Mr. David Ayres, of Camilla, Mitchell County, Southern Georgia—where snow never falls and the ground seldom freezes, and where the original pine forest is carpeted with grass that is indigenious to the soil, from January to December— says his sheep 3,500 in number, cost him annually 14 cents per head.   The average clip is three pounds of unwashed wool, which sells at 30 cents per pound, giving a clear profit of 90 per cent. on the money invested in sheep.   Mr. Ayres does not feed his sheep at any time, and relies entirely on native sheep.

Mr. John McDowell, of Washington County, Pennsylvania, keeps 670 highly improved sheep, the keeping of which costs annually $1.54 a head. He aims to make his wool clip clear which averages four pounds of brook washed wool to the sheep.   His wool crop sold in 1875 for

56 cents per pound, or $2.24 for each sheep sheared, but the crop cost, on account of the severe winter, 15c. a pound, which makes his net income per sheep $1.64. His sheep are worth $3.50 per head, and his net profits are 46 per cent. on the money invested. The land on which Mr. McDowell pastures his sheep is worth about $50 per acre, while I am authorized to sell land like that owned by Mr. Ayres for $1.50 per acre, and the 20,000 acres thus offered are situated in the same part of the State as that which sustains Mr. Ayres' sheep.

In Middle Georgia Mr. Robert C. Humber, of Putnam County, reports that he keeps 138 sheep of the cross between the Merino and the common stock. They yield an average of three pounds of wool per head, which he sold in 1875 at 25 cents per pound. They cost him nothing except the shearing, and he claims they pay him 100 per cent. on the investment, in mutton, lambs and wool. His sheep range on Bermuda grass fields in Summer, and the "old fields" in Winter. The "cane bottoms" and the cane which fringes with luxuriant leafage the streams in Georgia, offer stock of all descriptions most healthy and nourishing green food, even in the depth of Winter. No shelter is required, and no diseases of consequence are reported among the flocks of Georgia.

The arid wastes of Colorado, New Mexico and other parts of the "Great West," cannot be compared with the sunny skies and healthful climate and cheap lands, permeated by railroads and navigable streams, of Georgia. No treeless wastes or waterless tracts embarass the husbandman there, and Northern farmers are wanted by thousands to aid in developing the wonderful resources that are now dormant for the want of sturdy hands and intelligent labor that will create the capital needed. They will find a law-abiding people and a country so healthy that the death rate is less than in Maine, Connecticut or Missouri, and the same as in Michigan, which is conceded to be one of the healthiest States in the Union. Indented with some of the finest harbors on the Atlantic Coast, its transportation facilities are unequaled in the South. The whole State lies in the Temperate Zone. The price of land is from $1 to $50 per acre. The cheapest and best timber in the United States is there; no laborer was ever charged for fuel. The meadow in Southern Georgia is made by nature; the laborer can work in the field every day in the year.*

GEORGIA AND MINNESOTA COMPARED.

As an evidence that the true solution of the labor problem caused by the immense immigration to the Northwest *is to go South*, the following letter, recently received is here copied:

---

* I am indebted to the Report of the Department of Agriculture of Georgia, for the year 1875, for many of the facts stated above.

"PLEASANT PRAIRIE, Martin Co., Minn., }
                        MAY 13, 1880. }

"DEAR SIR:—I and several others of this place are in the sheep
and cattle business in a small way. Some of us got homesteads and
some bought slightly improved land for from $200 to 600 per quarter
section (160 acres). There is an abundance of grass here for hay and
pasture, thousands of acres of State, railroad and speculators' land
that we can pasture and mow free. It is ten miles to a railroad depot,
and good schools and citizens, *but no wood or timber*—(italics are
mine). *I have fed my sheep and cattle the past Winter just six
months. It takes three to four months to put up hay and feed for the
Winter here.* The wolves and dogs take on an average about ten
sheep a year from me. I keep about 500 sheep, medium quality, and
from 30 to 50 head of cattle, and my neighbors farm in a similar
manner. Now, what we would like is, to go to a place that is better
than this in some respects. We do not want to cut and stack hay
half of the Summer and fodder it out all Winter, which is nine months'
work, cutting, stacking and feeding.

                                        "GEORGE F. P."

I am prepared to sell immigrants as good land as that owned by
Mr. David Ayres of Camilia, Georgia, (who owns the 3,500 sheep referred to above), at from $1.50 to $3.00 per acre. I offer farms of 40
acres, with building for $225. I offer 1,000 acres to any one who will
make sheep husbandry a business and settle on the land for $1,000
cash. The owner has 3,000 acres and 1,000 sheep, which are not for sale.

The following letter from the most successful and enterprising wool-
grower and stock-raiser in Georgia is here appended:

                        ATLANTA, GA., May 24th, 1880.

FRANCIS FONTAINE, ESQ.,
        Commissioner Land and Immigration,

MY DEAR SIR:

I duly received yours of 19th inst., and herein very willingly answer
your questions.

At the South, merino sheep are seldom sheltered, and are fed but
little hay during the Winter; the gross weight of their fleeces, after
the hard washing spring rains, and freedom from hay seed do not
average with the fleeces of the Western merinos, stabled and high fed
as they are, and the rams often covered and blanketed, but the net
weights of the scoured fleeces of each section would show but little
variation.

As to the price of mountain land, no one can to this question make
a satisfactory reply. It is customary to use the pasturage of the
mountain ranges in common, and it would not pay to fence in large
tracts of such land, as the pasturage is only of value about five months
in the year, say 1st of May to 1st of October.

                Respectfully yours,
                        RICHARD PETERS.

The above reference to pasturage and mountain land, refers only to Northern and Middle Georgia.   The pasturage in Southern Georgia is perennial and good during the whole year.

The following letter is from one of the most respected and widely-known wool dealers in New York.   Mr. Lynch is also one of the Commissioners of Emigration for New York, being President of the Irish Society :

NEW YORK, May 24th, 1880.

FRANCIS FONTAINE, ESQ.,

160 East 10th St., N. Y.

DEAR SIR :

In reply to your favor of 19th, I beg to say that wool grown in Georgia is quite as valuable as the scoured wool of like quality grown in Indiana.   In both States the quality is nearly the same ; say $\frac{1}{4}$ to $\frac{1}{2}$ blood merino or "low medium ;" strictly fine merino's—" full blood "—have not done well in Georgia.   The sheep best adapted to the State are those of the quality and kind they have had there since the days of Oglethorpe.

The circular to which you refer was issued when Castle Garden was crowded with applicants for places and employers few ; now the case is reversed ; there are more orders for labor than can be filled.

Yours truly,

JAMES LYNCH.

The *Americus Republican*, of the 19th of May, 1880, gave an account of the sales of wool from a small flock of 13 sheep.   The sheep had yielded $13 worth of wool and 13 sheep.   The *Thomasville Enterprise* of July 15th, also states, that Mr. J. C. Lewis, of Thomas County, sold this year from 13 sheep 53 pounds of wool at 28 cents, making $14.85, and he reports 17 lambs.   He also sold 9 wethers at $2 per head.   From these sales it will be seen that the price of wool has declined, but that the profit at 28 cents a pound is quite satisfactory.

NOTES ON SHEEP HUSBANDRY.

The wool trade of Albany, Georgia, is assuming great value. About 500,000 pounds of wool are sold at this town per annum. Last year this trade in wool amounted to over one million of dollars in Southwest Georgia. There is nothing that would pay better than sheep raising, yet, strange to say, there are not more than from 800,000 to 900,000 sheep in the whole State, when we should have for wool and food 8,000,000 or more.

Mr. State Montgomery sold in Geneva, Talbot County, last year, $445 of wool from seventy sheep.

A writer in the Atlanta, (Ga.,) *Constitution* states that:

In 1871 he "bought 400 acres of reputed poor land in Glynn County, Georgia, and put upon it 100 sheep. In 1873, by natural increase, he had 376 ewes, and had sent to market 73 wethers. His sheep were penned nightly, and every two nights manured a half acre well. Since that time he had brought into a high state of cultivation 100 acres of land that seven years ago were considered worthless. Since 1871 he has bought 200 sheep, and now owns 1,800 head. He keeps a shepherd who is paid to attend to his business, and keeps an accurate book account of every dollar and dime spent on account of the sheep, and finds, by casting up a balance sheet, that it costs him exactly 57 cents a head per annum to keep his flock. They average him about three and a half pounds of wool each. Last year he clipped in May, and again in September, and the clip amounted to five and a half pounds per head. Last year he sold in Savannah and Macon 8,000 pounds of wool, at an average of 30 cents per pound, including a few pounds of merino wool, which makes the gross receipts $2,970. The annual expense of the flock $1,026. So there was an absolute profit in the wool of $1,664. Last year he sold in the above cities 92 wethers as mutton for $342, making a total of $2,289. Besides this, he has fertilized 84 acres of poor wire-grass land, so that last year he made from 10 acres in sugar cane 56 barrels of syrup, and 15 acres sown in oats yielded an average of 42 bushels to the acre.

The *Middle Georgia Times* says:

"We congratulate Georgia wool growers on their good fortunes in the advance price of that important commodity. It is said buyers are anxious to buy at forty cents per pound."

The average shrinkage of unwashed merino fleeces is from 48 to 52 per cent—about 50—and of unwashed (rams' fleeces), about 67 to 70 per cent. This would give as the shrinkage in ordinary washing from 17 to 20 per cent. The well known firms of Messrs David Scull, Jr. & Bro., and Coates Bros. of Philadelphia, and Mr. W. M. Brown, of Beverly Woolen Mills, endorse this statement.

The Waco (Texas), *Examiner* of May 25th, 1880, states, that at a meeting of the wool growers of Southwestern Texas, the following facts were elicited:

"Medium wools were quoted in the North at from 35 to 40 cents, coarse, 31 to 35, and Mexican 21 to 25. For wool worth 37 cents in the East, not more than 27 or 27½ could be obtained here. In Georgia, at the same date, wool brought from 35 to 40 cents a pound."

THE ANGORA GOAT IN GEORGIA.

This industry is peculiarly suited to all the Appalachian range from Virginia to Georgia, but no American State has demonstrated its truth as satisfactorily as has Georgia.   The wrong location of flocks is fatal to success.   If our farmers in Northern Georgia knew that this fibre for the last fifteen years has sold at double the price of the best combing wool, they would interest themselves more generally in providing an American growth of mohair.

ANGORA GOAT WOOL.

Statements have been going the rounds of the papers that there was no market for Angora wool.   The following is the testimony of Mr. R. Peters, who has a flock of 160 Angoras, which we give for the benefit of those who breed or intend breeding this valuable animal:

Q. What is the clip of the Angora goat?

A. From three to four pounds—the former for ewes and the latter for bucks.

Q. Do you find any difficulty in finding a market for it.

A. None in the world.   Messrs. Turner & Son, at Acron, Ohio, will take all that is produced at from seventy to eighty cents per pound for thoroughbred, fifty-five cents for full bloods, and forty cents for grades.

Q. What do you mean by full bloods?

A. The fifth cross, or those having 31-32 pure blood are classed as full bloods.

Q. What is the cost of keeping them?

A. Almost nothing.   They cost very little more to keep than the common goat, less than sheep, while they sell much higher and yield about the same annual income in wool as good sheep which clip an average of six pounds of wool.

Q. Have they been subject to disease of any kind?

A. None at all.   They have been healthy and prolific.   A late lamb is worthless, while if from any cause a kid is dropped late it makes as good an animal as those dropped early.

Q. You are, then, still pleased with them as profitable stock?

A. So well, so that I have made arrangements to make new importations with which to still further-improve those I already have, and which I am constantly improving by selection.

The following from the *Boston Advertiser* will show that Mr. Peters has carried out his intention:

"There are now in this city some fine specimens of a breed of Angoras never before, save in one case, exported from Turkey. The animals now under consideration arrived here a day or two since, in the steamer Dorian, from Constantinople, and were imported by C. W. Jenks. They are to form a part of the famous flock of Mr. Peters, in Georgia. They were brought some hundreds of miles on mule back to the coast from the province Geredeh, in the interior of Asia Minor. The Angoras heretofore received in this county, have been from provinces near the coast, and are smaller, with fleeces of four, five and six pounds. The Geredeh breed is larger, with fleeces eight, ten, twelve, and in some cases, fifteen pounds in weight, of very fine and silky mohair, a lock of which lies before us, with photographs of animals of this breed. Mr. Jenks has informed us that he has traversed hundreds of miles in the Blue Ridge mountains of North Carolina and Georgia, the altitude, climate and vegetation of which are a transcript of those of the region whence these goats were brought."

The *New York "South,"* in alluding to this importation, says:

"Fifteen years ago the mohair clip of the Cape of Good Hope had a value of $1,650. This year its value is $650,000."

Since the first importation, thirty years ago, the object of all the breeders in the United States seems to have been the raising of animals for breeding and not in flocks for fleece. There have been one or two exceptions; notably that of Mr. Richard Peters, who, on his fine stock ranche, among the foot-hills of the Blue Ridge, at Calhoun, in Northwestern Georgia, has, from the original stock, maintained a flock of greater or less number in their original purity.

No. 120.—Quantities of WOOL produced, imported, exported, and retained for consumption in the United States, from 1870 to 1879, inclusive.

Year ended June 30.	Production. a	Imports.	Total production and imports.	Exports.			Retained for home consumption.
				Domest'c	Foreign.	Total.	
	Pounds.	Pounds.	Pounds.	Pounds.	Pounds.	Pounds.	Pounds.
1870..	162,000,000	49,230,199	211,230,199	152,882	1,710,053	1,862,945	209,367,254
1871..	160,000,000	68,058,028	228,058,028	25,195	1,305,311	1,330,506	226,727,522
1872..	150,000,000	122,256,499	272,256,499	140,515	2,266,393	2,406,908	269,849,591
1873..	158,000,000	85,496,049	243,496,049	75,129	7,040,386	7,115,515	236,380,534
1874..	170,000,000	42,939,541	212,939,541	319,600	6,816,157	7,135,757	205,803,784
1875..	181,000,000	54,901,760	235,901,760	178,034	3,567,627	3,745,661	232,156,099
1876..	192,000,000	44,642,836	236,642,836	104,768	1,518,426	1,623,194	235,019,652
1877..	200,000,000	42,171,192	242,171,192	79,599	3,088,957	3,168,556	239,002,636
1878..	207,000,000	48,449,079	255,449,079	347,854	5,952,221	6,300,075	249,149,004
1879..	211,000,000	39,005,155	250,005,155	60,784	4,104,616	4,165,400	245,839,755

a In the column of "Production," the amount placed opposite the fiscal year is the production of the preceding calender year.

FRUIT GROWING AND TRUCK GARDENING.

Though the business is yet in its infancy, the shipment of early fruits and vegetables from Georgia this year will probably exceed 100,000 boxes. The Southern counties of Georgia are much more favorable to strawberry and kindred industries than Norfolk, Va., not being subject to the cold winds that are so damaging to Virginia, beisdes being further South, and hence, much earlier, which is the most important consideration to "truckers," as it means high prices. In fertility, transportation facilities, and variety of soils and productions, Georgia has advantages over any of the Atlantic Coast States, while taxation is lower than in any of them, being only 70 cents on $100 worth of property. Its railway communications give to its farmers speedy access to the Western and Northwestern markets, and refrigerator cars and first-class steamships convey the perishable products rapidly and securely to the Eastern markets.

Most of the early fruits and vegetables go to New York and are shipped by rail and water, at very low rates and on fast trains. The rate from Jacksonville to New York, by water, is thirty cents a box, and by rail 50 cents, with a difference in time of twenty-four hours in favor of the rail route. .

The shipment of strawberries begin in December and continue all through the season. They are sent to New York in refrigerators. In March of this years strawberries which had formerly brought $1 or $1.25 per quart were selling as low as 25 cents. The Western markets are now better than the Eastern. Georgia is exceptionally favored in its transportation facilities to Louisville and the Northwest. Tomatoes that bring $5.00 per bushel in Louisville, sell in New York for $2.

"Georgia has a monopoly of the early peach market in the great Eastern cities, and brings her large summer varieties into competition with the small early varieties grown in New Jersey. Southern and Middle Georgia are best adapted to the peach and fig. In these sections, lands along the lines of railways suitable for fruit-growing can be purchased according to quality and location, at from one to ten dollars per acre; but such lands are rapidly advancing in price. North Georgia is best adapted to the apple, and Middle Georgia to to the pear."

Mr. John H. Parnell, of West Point, Georgia, has probably the largest peach orchard in the United States. There are many large orchards in Georgia, and the report of the State Commissioner of Agriculture, states that the increase in the area devoted to fruits and vegetables in the past twelve months, is twelve per cent. Mr. Parnell shipped his crates of peaches to New York in May, and the first shipment to Atlanta, Georgia, was sold in that city for $35 a bushel.

Dr. S. Hape, of Atlanta, and Messrs. R. E. and H. A. Crittendon, of Randolph country, are now planting at Ward Station in that county probably the largest orchard and vineyard in the State. By another season they expect to have four hundred acres in trees and vines,

The "*Farmer and Fruit Grower*" expresses itself thus:

"If we were to make a prophecy about anything horticultural, it would be that, for years to come, the highest priced fruits to be seen in Western markets would be choice peaches."

Experienced fruit growers and gardeners can easily obtain lucrative employment in Georgia. By adding a little capital to their time and labor, they can affect most advantageous terms with the owners of the farms devoted to fruits and vegetables. A few instances will show how remunerative the business is, even when confined to local ma.kets.

"Mr. J. N. Walker of this county, shipped the past week to Atlanta, 377 cabbages, weighing about 3,500 pounds, which at four cents per pound, amounted to $140; nearly 40 cents apiece. One hundred and fifty of the above cabbage weighed 1,875 pounds, and brought $75, or 50 cents apiece."—*Brunswick Advertiser.*

On the subject of "Truck farming—What five acres will do," the Thomasville *Times,* says:

"Mr. Geo. P. McRae, of Lowndes County, planted two acres in cucumbers and three acres in tomatoes. He has or will ship six hundred crates of each, and for which two dollars per crate will be realized. This foots up the astonishing sum of two thousand four hundred dollars." The standard crate is 8 by 14 by 22 inches; capacity, one bushel. To insure ventilation they are generally made octagon shape, or corners of the headings cut off.

"Gen. LeDuc has in his possession several samples of Georgia raised tea, said to be of excellent quality, and for which the grower was offered fifty cents per pound in bulk for all he could supply. (Gen. LeDuc is U. S. Commissioner of Agriculture)."

Mr. Wm. Dean, a prominent woolen manufacturer and practical farmer of Wilmington, Delaware, has been traveling in Georgia and Florida:

(From the Jacksonville Union).

"Mr. Dean is enthusiastic in his description of what he has seen in the South. He thinks there is no comparison between the North and the South, so far as farming and cotton manufacturing is concerned, the advantages of the South being so much greater than those of the North. Very soon, he thinks, it will be impossible for the North to compete with the South in the manufacture of brown cotton goods. As for farming, Mr. Dean says, if he was thirty years younger, he could take two thousand dollars, go South, engage in farming, and in ten years have a hundred thousand dollars. He was glad to sell his potatoes at the North at a dollar and a half a barrel. It cost a half dollar to get them from where he lived to New York, and it cost no more to ship them from Georgia or Florida. Land in Delaware cost one hundred dollars an acre; here it can be bought for five dollars per acre. The latter, with judicious cultivation, would yield just as much as the former."

The *Albany News* states that Mr. J. W. Barnes picked fifty bushels of cucumbers from his truck garden just beyond the city limits. The most of them have been shipped to Philadelphia, where he expects to get four dollars per bushel for them.

Mr. Geo. T. Young, formerly of Plainfield, N. J., wrote from Waycross, Ware County, August 5, 1879, the following statement, showing how many crops can be raised from the same ground in the same year:

"We have gathered two crops of Irish potatoes, and have it planted now with sweet potatoes looking well. The longer I live here, the more I think of the country for agricultural purposes."

Hundreds of similar reports might be given, but the above statements are sufficient to prove that, whether early vegetables or fruit be cultivated, the area devoted to them need not be limited, as the demand in the Northeast and Northwest will always be in excess of the supply

Meanwhile it is well to note the effect of competition. The berry season North does not begin until June, and lasts only two weeks. Early fruits ripen in Southern Georgia before the seeds are planted at the North. For early vegetables and fruits, and rapid transportation to Northern markets, Southern Georgia is unexcelled. Georgia Strawberries are found in the New York market in mid-winter. But it is only in the past two or three years that the culture of strawberries has become a specialty in Georgia. The yield is said to be about 4,000 quarts to the acre, which, at 12 cents a quart, would give an income from 225 acres of over $108,000.

A Charleston newspaper notes the price of strawberries in South Carolina since they were first cultivated for Northern markets. In 1872, they brought an average price of 57 cents a quart; in 1873, 33 cents; in 1874, 38 cents; in 1875, 29¼ cents; in 1876, 21 cents; in 1877, 20 cents; in 1878, 11½ cents; in 1879, 14 cents; and this year the average is estimated at about 12½ cents. The decline in price is owing partly to competition with the West Indies and partly to the nominally low price of all products due to the return of the currency to a specie basis.

Good cultivators can make in Georgia an average of $150 an acre from strawberries above expenses. As much as $2500 *net* has been made from seven acres in strawberries in this latitude. Forty carloads of strawberries have been consumed in Chicago and sent to neighboring towns in one day, and Southern strawberries have been sold there when there was a foot of snow in Canada but a few degrees distant. The growers in Georgia having all the advantages which the early market gives, can make much greater profits than the growers in New Jersey.

Peas can be sown in Southern Georgia in January, and Irish potatoes may be planted at the same time. Beans can be planted in
February and March, and many other vegetables at same time.

The same advantage possessed by the coast of Georgia over that of
Virginia with regard to strawberries, pertain to the culture of early
potatoes for the Eastern markets. New potatoes from Norfolk, the
only Southern port quoted from ordinarily, begin to arrive in the
latter part of May or the first of June. They can be shipped from
Georgia several weeks earlier than from Norfolk, and the New York
markets are dependent on the Bermuda Islands, Florida, Georgia,
South Carolina and Virginia for early potatoes from January to July.
New York City and Brooklyn consume 10,000 barrels per day.

The receipts from sales of watermelons at Augusta, Georgia, are
worth $100,000 per annum. On the 20th April, a melon was
marketed at Americus, (Southwestern Ga.,) twelve inches long and
fourteen inches in circumference.

In the Spring of 1877, there were shipped from one port of Georgia
23,284 packages of fruit, 20,405 packages of vegetables, and 26,345
melons. I have not been able to get the returns for the year 1879,
but it is probable that there will be over 500,000 boxes shipped in the
year 1880.

The apples of Northern and Middle Georgia, keep well until Apri
or May. The Shockley ripens in September, and hardly ever fails
to yield a good crop. The Bachelor, Horn Apple or English Crab,
do well; the Nickajack or Fall Pippin are fine apples and ripen in
October. It is estimated that an orchard will yield from 50 to 75
barrels to the acre if well attended to. One-half will sell for $1.50,
and the other for $2 per barrel in New York, but there is a local
demand for all raised in Georgia at present.

The fruit farm should be near good shipping facilities, and there
should be an ample supply of labor in the immediate neighborhood
to furnish labor at any time needed. The coast counties of the State
of Georgia, with the quick sandy soil and sunny skies, can furnish
tons of ripe strawberries in mid-winter, and will eventually render the
growing of this fruit under glass unprofitable. In spite of the diffi-
culties that environ the Northern farmer, even the late crop of berries
are very profitable, and it needs but good judgment and business
foresight, to make much greater profits for the Georgia fruit grower.

The State Commissioner of Agriculture, in a recent report, shows
an increase in the area devoted to orchards and vineyards in the last
tewlve months to be twelve per cent. The most rapid increase will
be in peaches, grapes, pears and plums. Every part of Georgia has
grape growing enterprises in successful operation.

NOTES ON TRUCK FARMING.

OFFICE OF SAVANNAH, FLORIDA, AND WESTERN R. R., ?
AND FLORIDA DISPATCH LINE, 315 BROADWAY. ⟩

NEW YORK, May 11, 1880.

The receipts via " Florida Dispatch Line " and Southern Express Company for the week ending the 8th inst., were: Vegetables, 5.800 packages. Condition: cucumbers, mostly fair to poor; potatoes good; tomatoes good; beans poor; squash fair; cabbage fair; beets fair.

Prices: Potatoes, Florida, $2.50@5.00 per barrel; potatoes, Savannah, $3.00@4.00 per barrel; tomatoes, $2.50@4.50 per crate; cabbage, Savannah, $1.50@2.25 per barrel; cabbage, Norfolk, $1.00@2.25 per barrel; beans, flat, Florida, 50c.@75c. per crate; beans, round, Florida, 50c.@$1.00 per crate; beans, Savannah and Charleston, $2.00@2.50 per crate; squash, 50c. per crate; beets, $1.00@1.50 per crate; cucumbers, $1.00@1.50 per crate.

C. D. OWENS,

Gen'l Agent.

———

BOSTON, May 7, 1880.

Beans will not pay to ship any longer. Cucumbers plenty and poor; have ruled dull and lower; tomatoes, choice, sell fairly at quotations: Tomatoes, $3.00@4.00; cucumbers, 75c.@$1.59; beans, flat, $1.25@1.50; beans, round, $1.50@1.75.

———

PHILADELPHIA, May 15, 1880.

Potatoes, large to choice, $5.00@5.50; cabbage, $2.00@3.00 per barrel; tomatoes, ripe, $3.50@4.00; green, $1.00@2.00; cucumbers to-day, $2.50@3.00; beans and squash are not paying freight.

G. W. SHALLCROSS & CO.

———

FARNUM & CO., GEN'L PRODUCE AND COMMISSION ?
MERCHANTS, NOS. 108 & 110 FANEUIL HALL MARKET. ⟩

BOSTON, May 19th, 1880.

Editor Jacksonville Dispatch :

" The steamer United States arrived with cucumbers in fine order; they sold for $2.50 to $2.75 per crate, and found quick sale. Also the cucumbers via Atlantic Coast Line were in fine order, and sold at $2.50 to $2.72. We received twenty-six crates of cucumbers from G. R. McKee, of Valdosta, Georgia, via the steamer United States, which were the finest ever received in this market, and sold for good prices. FARNUM & CO.

GRAPE GROWING AND WINE MAKING.

Examine the isothermal lines and temperature tables of Georgia, and they will be found to correspond with the best wine growing countries in Europe. One thousand gallons of wine may be produced on an acre of land in Georgia, which can be bought for $5 per acre.

Any one familiar with the forests of Georgia has enjoyed the delicious flavor of the * "Scuppernong" and "Muscadine" grapes, even when in the wild state. They festoon the trees, and form an arbor, or support, with the aid of nature only.

White, in his description of this vine, says:

"We consider this very peculiar grape one of the greatest boons to the South. It has very little resemblance to any of the grapes of the other sorts. It is a rampant grower and requires little, if any, cultivation. It blooms from the fifteenth to the last of June, and ripens its fruit the last of September or beginning of October. It has no disease in wood, leaf or fruit, and rarely, if ever, fails to produce a heavy crop. We have never known it to fail. Neither birds nor insects ever attack the fruit.

"It will produce a greater weight of fruit than any other variety in the world. The clusters vary in size from ten to twenty berries, and the berries in size from three-fourths to one inch and a quarter in diameter.

"Vines, six years transplanted, have this year given an average of three bushels to each vine; we are credibly informed that a vine of this variety is growing near Mobile, which has produced two hundred and fifty bushels of grapes in a year, and we know that vines ten years old have given and will give thirty bushels per vine. A bushel of these grapes will give from three to three and a half gallons of juice, according to ripeness.

"It is the cheapest and most luscious of any grape we have ever seen or tasted, makes a fine, heavy, high-flavored, fruity wine, and is peculiarly adapted to making foaming wines."

(*From the Southern Enterprise, Agricultural Monthly, Atlanta, Ga.*)

"The grape, which promises to be, with the peach, the staple fruit product of the State, thrives well in every part of the State, but of course better in some localities than in others. Marked success has. been attained, however, in every section, showing the general adaptability of soil and climate to the growth of the vine.

"The Labrusca and Œstivalis types bear immense crops in every section. The Rotundifolia do not bear well on calcareous soils."

We give the names of some of the parties who have given especial attention to grape culture in different sections of the country:

---

* The original scuppernong vine, reported to have been discovered by Sir Walter Raleigh in the sixteenth century, covers over one acre of ground, and is assessed to produce over 2,000 gallons of wine in one season.

" Mr. John Stark, Thomasville, Georgia, has, for some years, been successfully engaged in grape growing and wine making. We know from personal observation that he has succeeded in making wines of fine quality.

"Messrs. Starowski and Schneider, Hawkinsville, Ga., have been remarkably successful, having produced from the vintage of 1878, twelve hundred gallons of wine on 2¼ acres of Concord and Delaware grapes. We sampled some of their Delaware wine, which was very fine at the age of one year.

" Mr. Jas. E. Antony, of Macon, Ga., exhibited at the last Horticultural Fair, grapes which challenged the admiration of all who saw them.

"Mr. W. W. Woodruff, of Griffin, Ga., has a large vineyard, from which he makes excellent wine.

" Mr. W. W. Clark, Covington, Ga., who has been somewhat a pioneer in this enterprise, had on exhibition at the Grape display of the Atlanta Pomological Society, last August, thirty-six varieties of grapes, all perfect developments of their varieties. He has gained quite a reputation for the excellence of his wines.

" Mrs. J. W. Bryan, Dillon, Walker county, in the Northwest corner of the State, on Lookout Mountain, has made a marked success of grape growing. So has Mr. J. Van Buren, a veteran fruit culturist near Clarksville, in Habersham county, in the Northeastern part of the State.

" Mr. P. J. Berckmans, of Augusta, though not making a specialty of grapes, has demonstrated that that part of the State is well adapted to their growth, while Mr. D. C. Schultze, near West Point, on the Western border of the State, has illustrated successful grape growing there.

" There are many fine vineyards around Atlanta and elsewhere, but we have mentioned enough to show how generally through the State the grape grows successfully.

"A company has been organized at Cuthbert, in Southwest Georgia, for the purpose of manufacturing wine on a large scale, and another is being organized in Atlanta, for the same purpose."

WATER POWERS AND MEDICINAL WATERS.

The difference of elevation between the sources among the mountains and the mouths of our rivers that empty into ocean or gulf, ranging from 1,000 to 3,000 feet, demonstrates the abundance and value of the water-powers of Georgia. We do not believe that Georgia is excelled in this respect by any State in the Union. In the appendix to this work will be found a partial list of the water-powers, of Georgia, but in order to illustrate briefly their extent and value we copy the following list from the official measurements of the State geologist:

"The estimates given below are for the theoretical horse-power of the stream, without the accumulation of its waters in a reservoir. The horse-power is equivalent to 33,000 foot-pounds.

	Horse-power.		Horse-power.
Chattahoochee river, Columbus	35,558	Coosawattee, Carter's mill....	3,085
Chatthoochee river, Fulton co.	2,448	Oconee river, Long Shoals F'y.	1,024
Ocmulgee river, Lloyd's Shoal.	3,970	Oconee river, Riley's Shoals....	2,054
Ocmulgee river, Seven Islands.	2,050	Oconee river, Oconee county...	5,642
Ocmulgee river, Clapp's Shoals	508	Oconee river, Jackson county..	271
Ocmulgee river, Glover's mill.	1,368	Tallulah river, Habersham co..	20,508
Etowah river, Bartow county..	2,250	Mulberry creek, Harris county.	1,020
Etowah river, Franklin mines.	1,029	Towaliga, High Falls..........	1,530
Etowah river, Lumpkin county.	272	Yellow river, Cedar Shoals....	1,302
Holt's Shoals, Bibb county...	1,050	Yellow river, Cedar and Henley	
South river, Butts county.....	350	Shoals..............	2,000
South river, Clarke's Factory..	247	Little river, Eatonton Factory.	155
Snake creek, Carroll county....	405	Nacoochee Gold Mining Co.,	
Pataula creek, Clay county.....	601	White county.........	575
Armuchee creek, Floyd county.	151	Savannah river, Augusta canal..	14,000

"These are only a few of the many which might be mentioned.

"The immense water-power of Anthony's Shoals, Broad river, in Wilkes and Elbert counties, has not been accurately estimated."

MINERAL WATERS.

Northern and Middle Georgia abound in Iron, Sulphur and Magnesia waters. Many of these have been for a long time attractive to invalids, and pleasure-seekers.

The following springs have been popular resorts for many years.

Catoosa Springs, Catoosa County.	Indian Springs, Butts County.
Rowland Springs, Bartow County.	Newnan Springs, Coweta County.
New Holland Spring, Hall County,	Sulphur Springs, Meriwether County.
Sulphur Spring, Hall County.	Warm Springs, Meriwether County.
Porter's Springs, Lumpkin County.	Chalybeate Springs, Meriwether Co.
Helicon Springs, Clarke County.	Gower's Spring, Hall County.

WARM SPRINGS, MERIWETHER COUNTY.

The situation, North side of Pine Mountain 1,800 feet above the sea, and natural advantages of these Springs are unsurpassed.

Except the Hot Springs of Arkansas, we do not know of any resort in America better suited to cure persons suffering from cutaneous diseases, rheumatism, dyspepsia, or diseases of the Urinary organs. The spring or fountain gushing forth from the mountain, 1,400 gallons of water per minute, temperature 90° Fahr., is one of nature's wonders.

The Baths, six in number, ten feet square, with a constant flow of water in and through them, from two to four and a half feet deep, are equal to any in America.   The analysis of the water is as follows:

Quantity ....1,400 gallons per minute.
Temperature... ....90° Fahr.
Specific Gravity.... 998-

IN THE WINE PINT ARE FOUND

Carbonic Acid Gas..................... ....1.11 cubic inches.
Carbonate of iron.......................... ....3.29 grains.
Oxide of Calcium.............................4.64   "
Oxide of Magnesia...........................11.68   "
Hydro Sulphuric Acid, a large quantity.

It is desired to form a stock company, and make of this estate an attractive Winter resort for Northern people who desire to escape the rigors of the Northern Winter.   There are 1,200 acres in this estate, all of which are admirably adapted to sheep husbandry and fruit culture.   There is a cold Chalbybeate Spring near the hotel, and the Chalybeate and White Sulphur Springs, both popular Summer resorts, are each distant seven miles.   The North and South Railroad is a few miles distant.   For particulars, address ·"Commissioner Land and Immigration, Atlanta, Georgia."

The best improved watering place in the State is the Catoosa Springs, Catoosa County.   There are twenty-four different varieties of mineral water at Catoosa Springs; the depot of the W. & A. R. R., is two miles distant.

The coldest freestone water is about a half mile from Dahlonega; it registers a temperature of 55° Fahr.

Porter's Springs in Lumpkin County, 28 miles north of Gainesville, and 10 from Dahlonega, is a very popular resort, surrounded by the finest scenery in the State.   The water is cool, and the temperature delightful during the entire season.   Their altitude above the sea is 3100 feet, affording the purest cold free stone as well as best mineral waters.

NORTHERN GEORGIA, TALLULAH FALLS, HABERSHAM COUNTY.

TYPICAL FALLS, NORTH-EASTERN GEORGIA.

If the mountains of North-eastern Georgia recall the groups in the "Saxon Switzerland," of Saxony, the Nacoochee Valley and the neighboring falls of Tallulah and Toccoa rival the most picturesque falls and cascades of Switzerland. The falls of the Sallenche, in the Canton Valais, sometimes called the Pissevache, is seen again in Toccoa, while the Tallulah Falls recall the Giesbach, or "Dust Fall." But, though resembling the mountains and cascades of the Middle Alps, the altitude of which is from 2,000 to 5,500 feet. Tallulah possesses characteristics that are distinctly American. But for the seething, rushing torrent, making its impetuous way through the narrow ledge walled by precipitous cliffs, Tallulah would be like the cañons of the far West. The Genevese, however, will see here a resemblance to the cliffs of the "Grand Gorge," on Mont Salève, near Geneva. The perpendicular heights of solid rock are thus named:

Point Inspection	1,200 feet.	Grand Chasm	800 feet.
Throne of Aeolus	600 "	Vulcan's Forge	500 "
Devil's Pulpit	450 "	Lover's Leap	500 "
Student's Rostrum	500 "	Diana's Rest	200 "

The altitude of the hotel is 2,382 feet above the sea. The principal falls are "Lodore," 46 feet high; "Tempesta," 81 feet high; "Hurricane," 92 feet high, and "Oceana," 49 feet high. The "Bridal Veil" fall, 26 feet in height, is the beginning of the mighty rapids of the Tallulah river, which, in a distance of three-fourths of a mile has a total fall of 450 feet. The height of the perpendicular cliffs, tufted with trees and shrubs here and there, on either side, is from two hundred to one thousand feet. Hawthorne's Pool, between Lodore and Tempesta Falls, is a lovely basin. Here a minister by the name of Hawthorne lost his life by venturing to take a bath in the seductive but dangerous waters. It is a thrilling scene to look down from the Grand Chasm at the foaming torrent, rushing with tempest haste in its narrow channel, eight hundred and sixty feet below. Distant fifteen miles away is the loveliest valley in Georgia— "Nacoochee"—named after the indian maiden who, the legend says, took the fatal leap. In this beautiful valley is the mansion of Mr. James Nicholls, surrounded by beautiful grounds, ornamented with flowers, rare plants, artificial lakes, fish pools, and parks for deer. Upon one of the pre-historic mounds that extend from Kentucky to Mexico—relics of the unknown race that first peopled this continent—is a fountain which is watered from the upper waters of the Chattahoochee river.

Toccoa, on the Air Line Railroad, is the nearest station to this region. It is probable that thousands of visitors would visit Tallulah

Falls each summer, if sufficient hotel accommodations were provided. For information concerning these properties, or for water-powers and manufacturing sites in any part of Georgia, address "Commissioner Land and Immigration, Atlanta, Georgia."

TOCCOA FALLS, NORTH-EASTERN GEORGIA.

THE CULTURE OF COTTON.

As cotton is the leading or "money crop" of Georgia, many may wish to know how to cultivate the crop of 600,000 bales, worth last year over $35,000,000.

THE COTTON PLANT.

The producers of Georgia numbered at the last census, 888,081 people.   The following brief description is offered:

In the rapid, loose, soil-denuding way of the *ante bellum* period, and on large plantations to-day, one mule and two "hands" (a hand means an adult laborer,) cultivated 30 acres in corn (maize) and cotton, and 20 acres in oats and wheat, additional help being necessary to harvest the crops and prepare them for market.

The Northern farmer, accustomed to careful tillage and manuring, will find the *intensive* system of farming as profitable in Georgia as in New England, and will reduce the area to be cultivated one-half or

three-fourths with profit.   The extra cost for manures and extra care
in tillage required to make a bale per acre, instead of a bale on three
acres, is more than compensated by the saving in labor, owing to the
smaller area cultivated.

The U. S. Agricultural Department, in its report for 1879, thus
compares 1878 and 1879:

	1878.	1879.
Cotton, bales	5,216,603	5,020,387
Cotton, price	$193,854,611	$231,060,000

Thus a few cents difference in the price of the raw material makes
an astonishing difference in the figures.

It is worthy of note that cotton sold at twelve cents per pound (the
price realized this year,) will not buy any more meat or guano than
the cotton which sold at eight and a half cents (8½ c.) per pound last
year.   A ton of guano bought a bale of cotton in the spring of 1879,
and it did the same thing last spring, although the price of cotton is
very much more in 1880.

There are four stages: First, planting; second, thinning to three
plants to a place, and then the second "chopping out," at an interval
of two weeks, when the plants are reduced to one, each plant being a
little more than the width of a hoe distant from the other; third, the
Summer growth; and fourth, the maturing and picking season.

In cultivating cotton, the farmer, for the first two years, will need
only a hoe, a bull tongue or "scooter" plow, and a "turn shovel
plow;" the whole cost for cotton culture implements not exceeding
$12.00.

In Southern Georgia cotton planting begins with April, is continued until June if necessary, and begins to mature in August, continu-
ing to ripen until frost, according to the time when the seed was
planted.   The cotton seed has a tap-root which strikes downward in
a week from the time of planting.   In a month the first "chopping"
begins, the "turn plow" having previously cleaned the land of grass
and leaving the cotton bed elevated in a narrow row, so that it is
easy work for the hoe to reduce the plants to a "stand," ie., two plants
to the hill.   ·

The seed is dropped in drills in the rows, which run at distances
varying from three to five feet, according to the fertility of the land—
the richer the soil, the wider the rows—for cotton is essentially a sun-
plant, and, on rich soil, the weed is too luxuriant to admit the sun's
rays, unless the "rows" are made wide.   In two or three weeks the
ground is again plowed; the hoes follow, eradicating all grass and
cutting away all the plants but one in a place.   At twelve inches the
plant begins to throw forth limbs on which appear leaves and buds,
the pure white blossom opening at sun rise or after, and closing early
in the afternoon, turning as it closes to a reddish color.

Little children and women can easily aid in cultivating and harvesting the cotton crop. Few crops present a more beautiful appearance to the eye than the cotton crop when in full bloom, resembling rather a flower garden than a field crop. Finally the flowers drop, leaving a litttle "boll," which, in about six weeks, matures its fruit of fleecy cotton—truly a "golden fleece."

There are three crops, known as "bottom," "middle" and "top" crops. The first is from blooms that come before the middle of July; the second is from blooms that come between July and September, and the last is from blooms coming after September 1st.

Just here we will note a fact that, while generally recognized, has not yet been accounted for, viz: Northern Georgia was not considered a cotton country until after emancipation; now Atlanta receives, with two exceptions, more cotton than any city in the State, and cotton is cultivated with profit among the mountains, even to the Tennessee line. This is probably due to two facts: First, Northern Georgia has always been a "white man's country," few negroes comparatively and few large plantations having existed there before the late war; second, the white laboring man in Georgia is generally an industrious laborer, and has too much intelligence to work large areas of poor land or to raise crops that do not pay. The best paying crop for the poor man who works his own land is cotton, because the product of twelve acres at one bale to three acres or of four acres at one bale per acre, worth in either case about $200, may be carried to market in a two-horse wagon at one journey. Any farmer can estimate the cost of transporting any other farm "money crop" of same value to market, and realize the advantage of the farmer in Georgia. Again, a handful of cotton seed to the "hill" of corn—applied before planting the seed—will generally double the crop of corn (maize,) and four bales of cotton, after being marketed, leave as manure, in addition to the stalk and limbs, 4,400 pounds of cotton seed. The chemical and manurial value of cotton seed could easily be given by comparative tables as proved by analyses, but we desire to make such illustrations as will be understood and appreciated by the plain farmer.

The cotton plant was originally a tree, or at least an herbaceous perennial, and still retains the tap-root, on the skilful management of which the crop very much depends. The problem, suggests a recent writer, is so to increase fine and lengthen the lint, that the seed cotton of the future will have 40 or 50 per cent. of seed, and not 65 or 70 as is the case now. There is no single crop more worthy of investigation, both from the scientific and practical standpoints, than that of cotton.

NOTES ON COTTON CULTURE.

The following tabular statements taken from the diary of a noted cotton planter, will give additional light to this subject:

1859–60—First bloom, May 31; killing frost, Nov. 7; total crop, 4,675,770; total value, $271,783,807.

1860–61—First bloom, May 25; killing frost, Oct. 30; total crop, 3,700,000; total value, $185,000,000.

1861–62—First bloom, May 31; killing frost, Oct. 13.

1862–65—[No data.]

1865–66—First bloom, June 23; killing frost, Oct. 20; total crop, 2,151,043; total value, 450,084,227.

1866–67—First bloom, June 11; killing frost, Oct. 25; total crop, 1,951,988; total value. $282,272,498.

1867–68—First bloom, June 1; killing frost, Nov. 6; total crop, 2,430,893; total value, $232,320,444.

1868–69—First bloom, June 11; killing frost, Oct. 28; total crop, 3,154,592; total value, $329,586,115.

1870–71—First bloom, June 9; killing frost, Nov. 18; total crop, 4,347,006; total value, $306,376,982.

1871–72—First bloom, June 4; killing frost, Nov. 15; total crop, 2,974,351; total value, $327,178,610.

From the above statements it will be seen that a small crop of cotton is worth as much to the producers as a large one, and, therefore, it is questionable whether low wages and large cotton crops represent a wise policy. The more numerous the small farmers, the greater the aggregate profit, seems the rational conclusion.

THE COTTON ACREAGE OF 1880.

The New York *Financial and Commercial Chronicle* has made up its figures of cotton acreage, stand and condition for 1880, and arrives at the following result as to acreage:

States.	Actual Acreage, 1879.	Increase.	Estimated for 1880. Acres 1880.
North Carolina	624,089	8 ⅌ ct.	674,016
South Carolina	985,370	11 ⅌ ct.	1,093,760
Georgia	1,744,048	10 ⅌ ct.	1,918,452
Florida	222,705	3 ⅌ ct.	229,386
Alabama	2,122,422	8 ⅌ ct.	2,292,245
Mississippi	2,117,101	3 ⅌ ct.	2,180,614
Louisiana	1,285,250	4 ⅌ ct.	1,336,660
Texas	1,684,631	17 ⅌ ct.	2,971,018
Arkansas	1,132,876	16 ⅌ ct.	1,314,147
Tennessee	761,400	15 ⅌ ct.	875,679
Total	12,679,962	9.51 ⅌ ct.	13,885,947

The average yield of lint cotton, per acre, is given as follows:

1870-1	pounds per acre,	191
1871-2	" "	147
1872-3	" "	177
1873-4	" "	169
1874-5	" "	154
1875-6	" "	177
1876-7	" "	171
1877-8	" "	169
1878-9	" "	182

PROFITS OF COTTON CULTURE.

A correspondent of the New York *Bulletin*, a planter himself, thus describes the profits of cotton culture:

"The average estimated cost of raising, ginning, baling and delivering the crop at the railroad is about $11 per acre, and the average yield of the South is 191 pounds per acre; that is to say, the cost of raising cotton is 5¾ cents per pound. The planters have received an average of about 11¼ cents for it delivered at the railroads, thus making a profit of about 5½ cents per pound on at least five million bales of 450 pounds each—2,250,000,000 pounds—or $124,000,000 clear profit."

NOTES ON COTTON CULTURE.

PRICES OF COTTON FROM 1825 TO 1878.

	Cents.		Cents.		Cents.
1825	27	1843	8	1861	28
1826	14	1844	9	1862	68
1827	12	1845	9	1863	69
1828	13	1846	9	1864	1.90
1829	11	1847	9	1865	1.22
1830	13	1848	8	1866	52
1831	11	1849	11	1867	36
1832	12	1850	14	1868	33
1833	17	1851	14	1869	35
1834	16	1852	10	1870	25¾
1835	20	1853	11	1871	25¼
1836	20	1854	10	1872	25⅝
1837	17	1855	11	1873	21⅞
1838	12	1856	12	1874	18¾
1839	16	1857	15	1875	17¼
1840	10	1858	13	1876	13½
1841	11	1859	12	1877	13 5/16
1842	9	1860	11	1878	13 3/16

The highest price for middling upland for the cotton year 1877-8 was reached August 31st, at 12 3/16 c., and the lowest, 10 9/10 c. The lowest price reached in the period of years given above was 4c., in 1845. In 1843, '44 and '48 the lowest figure reached was 5c. The next lowest was 6c., in 1846 and 1849, and 7c. in 1855, since when the lowest was attained December 13th, 1878, at 8 13/16 c. The highest price of middling upland in the cotton year 1878-9, has been 13¾ c.; this has also the distinction of being the highest quotation since 1875.

COTTON SEED OIL.

A letter from Boston says on this subject: "Cotton seed oil is now used but little in the paint trade, its use being confined mainly to the manufacture of putty. Where it was formerly used, linseed oil is now substituted to a great extent. It is, however, used quite largely in the preparation of woolen cloth and morocco leather, and for oiling machinery.

"It consists of three grades—the clarified, the refined and the winter bleached. The great use to which the refined and bleached oils are now put, is as a substitute for almond oil or genuine Italian olive oil. Great quantities are sent to Europe annually, where it is transformed into a sweet oil which it would puzzle any but a connoisseur to detect from the genuine article.

"This use to which cotton seed oil is now put, although of comparatively recent origin, is constantly growing in importance, and it will, ere long, become one of the most important items in the list of our exports to the old world.

"Thus the great staple, cotton, not only clothes our nakedness, nourishes our cattle in the form of cotton seed meal, but is now used to render our own food palatable. Who can venture to tell to what further uses it may yet be put?

"The market value of cotton seed oil depends to a great extent upon that of lard oil, as both articles are used for a great many similar purposes."

In Southern Europe, olive oil is now largely adulterated with oil of cotton seed. Forty-one mills for extracting oil from cotton seed are being worked in the cotton belt, and there is much money in this industry.

NOTES ON COTTON SEED MEAL.

The following table exhibits the relative feeding value of cotton seed and cotton seed meal:

	Digestible organic matter.	Albumin. oids.	Carbo. Hydrates.	Fats.
Cotton Seed	56.	17.1	11.6	27.3
Cotton Seed Meal	57.5	30.	17.5	10.

One hundred pounds of cotton seed are worth about as much for producing milk as the same quantity of cotton seed meal, and, therefore, cotton seed at 10 cents per bushel, is a much cheaper cow food than cotton seed meal at $20 per ton. A ton of cotton seed, (66 bushels to the ton) will cost $6.66, against $20 for the ton of meal.

For manurial purposes the meal is much more valuable than cotton seed, for the reason that the oil which has been expressed, being twenty per cent. of the whole weight, is worth nothing as a manure. For feeding purposes, however, either for milk cows or fattening animals, the oils, rich in animal fats, is very valuable.

ATLANTA ROLLING MILL.

IRON MANUFACTURE.

Mr. Robert P. Porter, of Chicago, one of the ablest and most widely known statisticians in the United States, in a letter to the New York *Sun*, thus writes:

"I made the statement that pig iron can be manufactured in the South more cheaply than in any other part of the United States cheerfully enough, and that Georgia was never in a better condition than to-day. I will now supplement it with the fact that at Knoxville, Chattanooga, Atlanta, Rome, and many other points in Kentucky, Tennessee, Georgia and Alabama, they have engaged successfully in the manufacture of iron in all its forms, drawing their raw materials from Southern mines of ore and coal and limestone, deposits of which are abundant and of the best quality. I believe that Northern capital as it gains confidence in the South, will, ere long, help it to open new mines of ore and coal, to build new furnaces, and to extend its railroad facilities.

"I fully agree with Mr. Fontaine when he says the South is a fine field for emigration. The abolition of slavery will make it less difficult for white emigrants to compete in agricultural pursuits."

Blast furnaces are numerous in North-western Georgia, some of them producing 40 tons pig iron per day, and many immense establishments, for the manufacture of pig metal, demonstrate that this industry will rapidly assume first-class importance in Georgia. "Charcoal iron" cannot be manufactured to better advantage anywhere in the world. The Rolling Mills at Atlanta have shipped immense quantities of rails to Texas and other States. In the region of country alluded to in Mr. Porter's letter, may be found coal, iron, limestone, and forest, placed by nature in near proximity to each other. River and rail transportation is convenient and accessible. The following extract from a letter received by the Commissioner of Immigration, explains the situation from the laborer's standpoint:

"Mr. W. wrote you a few days ago our proposition with regard to the fifty men needed at once, which was to advance funds for transportation to this place, (Cedartown, Cedar Valley, Georgia), give them permanent work at eighty cents per day, and furnish them houses, fuel, and space for garden, free of cost. We would prefer the number to be composed of as many single men as could be obtained, and would expect them to contract to work for the Company for a term not less than twelve months. We believe that a good class of emigrants would make a satisfactory substitute for the negro. We enclose pass for Thomas Graeme, Scotch emigrant, over our road to this place, and check for his transportation from New York to Cartersville. The cost of transportation will be deducted from the wages of all those to whom we advance money."

Yours, truly,

C. Iron & R. R. CO.,

J. R. B., *Sec'y*.

LIST OF IRON FURNACES IN GEORGIA.

					Capacity Tons per Day.
1. Bartow Furnace,	Bartow Station,	Bartow Co.			20
2. Charcoal "	" "	"	"		7
3. Rogers "	Rogers "	"	"		7
4. Pool's "	Stamp Creek	"	"		4
5. Brown and Thomas Furnace,	" "	"	"		4
6. Cherokee Furnace,	Polk	"			40
7. Ætna "	"	"			10
8. Ridge Valley Furnace,	Floyd	"			12
9. Rising Fawn,	Dade	"			50
10. Ward's Diamond Furnace,	Bartow	"			4
11. Stamp Creek Furnace,	"	"			4
12. Etowah Furnace,	"	"			4
13. Allatoona "	"	"			4
14. Phœnix "	Dade	"			40
15. Cherokee "	"	"			40
					——
					250

Manganese with Oxygen forms Pyrolusite, from which Ferro-Manganese is made in Bartow County, containing 60 per cent. of Manganese, worth $180 per ton.

For smelting copper, says the Hand Book of Georgia, there were, before the war, extensive works erected at the "Mobile Mine" in Fannin County, but they were burned and have not been rebuilt.

The "Hiwassee Mine," in Towns County, will probably be worked again soon. At the "Waldrop Mine," in Haralson County, the Tallapoosa Mining Company have cut a vein of chalcopyrite, etc., yielding, on an average, 8 per cent. for 125 feet longitudinally, in a drift that has been opened, and the bed of the ore found to average five feet in thickness for this distance. It is from 80 to 100 feet from the surface.

No. 128.—Quantity of PIG-IRON produced, exported and retained for consumption in the United States, from 1872 to 1879, inclusive.

[Expressed in tons of 2,240 pounds.]

Year Ended June 30.	Production.	Imports.	Total production and imports.	Exports (foreign and domestic.)	Retained for home consumption.
	Tons.	Tons.	Tons.	Tons.	Tons.
1872	1,706,793	247,529	1,954,322	2,712	1,952,150
1873	2,548,713	215,496	2,764,209	2,818	2,761,391
1874	2,560,963	92,042	2,653,005	10,152	2,642,853
1875	2,401,262	53,437	2,454,699	16,193	2,438,506
1876	2,023,723	79,455	2,103,188	7,241	2,095,947
1877	1,868,961	67,922	1,936,883	3,560	1,933,323
1878	2,066,594	55,000	2,121,594	6,198	2,115,396
1879	2,301,215	87,576	2,388,791	3,221	2,385,570

THE MANUFACTURE OF COTTON YARNS.

Georgia and the Carolinas furnish the Philadelphia weavers with a considerable portion of the cotton yarns that they use.

The following estimate is believed to be correct. It is based upon present prices of cotton and yarn, which are very favorable to spinners. One system consists of—

576 spindles, at $15 per spindle	$8,640
Engine and boilers	2,500
Buildings and grounds	2,500
Shafting and belting	1,360

Total cost ...., ... $15,000

Producing—

120 bunches of yarn, per day, 5 lbs. @ 28c., $1.50	$138.00
Cost of the cotton, 6 lbs. @ 12½c., 75c	90.00

Put in the market and sold—

Freights, commissions, etc., 10½c. per bunch	12.60
Cost to make per bunch, 11½c	13.80

Total cost of 120 bunches $116.40

Deduct this from the value of the 120 bunches and we have—

Net product per day	21.60
" " " month	561.60
" " " year	$6,739.80

This calculation gives a profit of more than 40 per cent. on the capital invested.

―――――

CLEMENT ATTACHMENT MILLS.

From the report of C. T. Harden, Manager, Windsor, N. C., who has one of the mills in operation:

"We started last June, and have been running smoothly ever since. We are pleased with our mill and have already enlarged it, and are going to enlarge it to double the size it is in the Fall. We are now running two attachments, 512 spindles. Our mill cost $11,000 as it now stands. We are averaging 300 pounds of first-class yarn per day. Our mill is paying 35 per cent. on the investment, and we expect to make it pay 45 per cent. as soon as our hands become expert. We have not got a hand that ever saw a mill before. We have met with no reverse, and had no mishap to stop the mill a day since starting. There is an unlimited demand for our yarns. We get the highest market price for our goods.

"Respectfully,

"C. T. HARDEN."

It is stated that under the old system it cost $5 a bale for ginning, baling, &c., or about one cent per pound. We now quote:

"Mr. Webber in his work (Manual of Power, page 97,) gives the expense of taking baled cotton through the opener, picker and card ing machines at the Boot Cotton Mill, at 66.100 per pound. On that plan, accordingly, it would cost $1,166 to make 100 pounds of sliver from seed cotton. On the Clement plan the top flat card will card as much as four on the old plan (for reasons see our illustrated articles on the Clement Attachment.) Three Clement machines connected by railway will make 600 pounds of sliver per day, and when the self-opera-ting feed rigging (the Bramwell Wool Feeder) is applied to the Clement machine for a feeder, one man can attend three machines, which, to include all expenses, will not amount to more than $2 per day, or 33 cents per 100 pounds of sliver made. This will amount to a saving of about $1.33 on each 100 pounds of cotton.

"This estimate on the expense of ginning is based upon the usual practice (where gins are owned by other than the planters or cotton manufacturers,) where parties own gins and gin cotton on shares or for money. Even if $1 is saved on the hundred pounds of cotton it would amount to a very large saving. Take a mill with the capacity of the Boot Cotton Mills of Lowell, that manufacture 172,000 pounds of cotton per week, this would give them an additional profit of $1,720 per week, or $89,440 per year, or 7.45 per cent. on their capital stock of $1,200,000. This is as cotton is handled at present."

In answer to enquiries as to the cost of the Clement Attachment, Mr. E. E. Whitefield, Sr., writes to the *Planter's Journal*, stating that it is difficult to give the exact cost of each piece of machinery.

"He says, however, that the cleanser of seed cotton costs $75, and will cleanse enough cotton for ten cards and attachments. The attach-ment, with feed table, chute traverse and stop motion, costs at the Memphis shop, $300, and the small drawing and cam motion costs $75. The royalty on each attachment is at present $150. •

"A one-card mill can be driven by a seven-horse power engine, and requires seven operatives, all women and girls, to convert 650 to 675 pounds of seed cotton per day into 190 to 195 pounds of yarn.

"It is estimated that every attachment will require $5,000 invested in permanent machinery, houses, etc., and $5,000 for working capi-tal. To convert 120 bales into yarn would therefore require an outlay of $10,000."

Several Clement Attachment mills will be built in Georgia this year. Where there is water power not needed for other purposes the investment is reduced.

The Westminster (S. C.) Clement Attachment Company, composed of farmers, cleared last year 40 per cent. on the investment, though they run in great part with second-hand machinery.

JUTE BAGGING MANUFACTURE.

Jute bagging, being used chiefly for covering cotton, should certainly be manufactured in the cotton States; there are but two in existence in the South, one of which is at Columbus, Georgia, and their manufactured products command immediate sale at good prices. The following report of the Committee appointed by the New Orleans Cotton Exchange to enquire into the manufacture of jute, will be read with general interest :

(Report on the Manufacture of Jute Bagging.)

A Committee appointed by a meeting held recently at the New Orleans Cotton Exchange, to inquire into the manufacture of jute, have reported as follows :

"Your Committee have ascertained that fifty-five thousand ($55,000) dollars will cover the entire outfit of a jute bagging mill of twenty-seven looms, all of most approved machinery, and erected in mill building in this city.

"The party whose estimates we have adopted, Mr. S. D. Randall, is a lifelong bagging manufacturer, and himself announces a desire to have his name put down for five thousand ($5,000) dollars of the capital stock, and also tenders his valuable service for superintending the mill, and at a very moderate salary.

"He guarantees the product of the mill, when well under way, to exceed thirteen thousand (13,000) yards per day, of twelve (12) hours.

"With an annual production of, say, only 3,900,000 yards, the saving of transportation charges from New York—⅝c. per yard— would of itself yield a net profit to the mill of $24,375, as raw material, fuel, labor and supplies are as low, or lower, here than there.

"If we take the present cost of two pound jute bagging in New York, of 10¾c. (11c. per yard,) we find that 100 yards would cost, brought here from New York, $11.37. We could buy jute rolls in New York, pay freight on them and deliver at 3½c. per pound, making the entire cost of 100 yards manufactured by the mill here, not to exceed $9.12, showing a profit on 100 yards to the mill, of $2.25.

"H. DUDLEY COLEMAN,

" Chairman of the Committee."

Columbus is the fifth town in population in Georgia. Four railroads terminate there, and it is situated at the head of navigation of the Chattahoochee river, which is navigable to the Gulf of Mexico, a distance of four hundred miles. The society is refined, hospitable and intelligent; the public and private schools being equal in efficiency to those of any town in the United States. The Columbus Female College is in a prosperous condition. Columbus and its suburbs was formerly the home of many of the wealthiest planters in the State. Its commercial importance is great, being the chief market of Eastern Alabama and the surrounding counties in Georgia.

Columbus receives about 80,000 bales of cotton annually. Selling directly to merchants, the Eagle and Phoenix Mills, of Columbus, have now a trade extending from Eagle Pass to Richmond, Va., and from Fremont, Nebraska, to Milwaukee, Wisconsin.

The water-power of the Chattahoochee river alone, at Columbus, Ga.—the Lowell of the South—is unequalled as an investment for manufacturing cotton, all things considered, in the United States.

Commencing with none in 1866, Columbus now operates 60,000 spindles and 2,000 looms, besides many other industries. The various manufactories here give steady employment to 1,201 men, 1,160 women, and 280 children : total, 2,801. We do not count the colored laborers. The cotton mills have taken from September 1st, to May 7th, 15,618 bales; 3,000 more than last season. This cotton at present rate of low middling, would bring $819,945. The planters and merchants would get this, and a few others a little more, were we without the factories. This cotton, however, went through the mills.

713 Looms, 20,300 Cotton Spindles, and 2,300 Woolen Spindles. 1877.

and its value increased three-fold, brings $2,459,845, making the State richer $1,600,000. On some classes of finer goods manufactured here values are greatly enhanced. The last fiscal year the factories reported sales from their offices in the city at $1,417,722. This does not include a large mill outside the city, and the sales by agents abroad.

Gen. Lee surrendered at Appomattox, April 9th, 1865. Gen. Wilson, of the United States' Army, burned at Columbus, ten days after the surrender of Gen. Lee's Army, in addition to all the factories, foundries and industrial establishments, 60,000 bales of cotton, worth $1.22 per pound, the average weight being 500 pounds to the bale. The aggregate loss was over $20,000,000, and the inhabitants were reduced from affluence to want. The finest water power in the Southern States, and one of the finest on the continent, was at their doors. The Chattahooche River at Columbus, has 125 *feet fall in a distance of* 2½ *miles*. This water-power at the head of the Chattahooche, was then, and is still, owned by citizens of Columbus, who offer it to manufacturers at exceedingly low prices. Besides the largest Iron works South of Richmond, Virginia, there are many enterprising establishments, among them a mill for manufacturing Jute Bagging. There is but one other jute factory in the Southern States. Extensive plow and iron works, wood establishments, flouring mills, machines for the manufacture of ice,* and many other industries, all erected on the ashes created by war, and put in operation since 1867 with home capital. These are the forcible arguments which Columbus presents in favor of Southern manufactures.

For water powers and manufacturing sites address " Commissioner Land and Immigration," 60 East 10th Street, New York, or Atlanta, Ga.

---

*Ice is sold in Georgia at the rate of 8 pounds for five cents, while in New York City it is priced at 70 cents per 100 lbs., and it has cost in New Jersey this year $2 per 100 lbs. Lake ice is sold in Columbus, Georgia, at 75 cents per 100 lbs.

EAGLE AND PHŒNIX MILLS, COLUMBUS, GEORGIA, 1880.

Eighteen hundred operatives, forty-five thousand spindles, and eighteen hundred looms, manufacturing over one hundred varieties of white and colored goods, using thirteen thousand bales of cotton annually, and eight hundred pounds of washed wool daily.

COTTON MANUFACTURE IN GEORGIA.

The number of spindles in the Southern States are reported as follows :

States.	Spindles.
Arkansas	1,700
Alabama	63,000
Georgia	213,000
Kentucky	11,264
Louisiana	6,200
Mississippi	79.000
Maryland	113,000
Missouri	26,000
North Carolina	93,300
South Carolina	92.000
Texas	9,300
Tennessee	49,500
Virginia	52,000
Total	713,200

It will be noted that Georgia is far ahead of any of her sister Southern States and of the 213,000 spindles in operation, Columbus has 60,000, and all put in since 1866.

Of the population in Georgia in 1870, only 6 per cent. were engaged in manufacturing. The census of 1880 will show material progress in this respect, for cotton manufacturing is the most profitable industry in the State. Of the whole population of Georgia in 1870, only 11,127 were foreign born, and the operatives in the factories, except the foremen, are natives. No negro is employed in any cotton factory, and "strikes" are as yet an unknown factor in the labor system of Georgia. The white natives learn rapidly and readily. Labor, being better paid in factories than in agriculture, is easily obtained, though the wages paid are less than in New England. The laborers, who live generally in villages with ample ground for gardens, and houses furnished rent free, seem healthy, contented, and cheerful.

The Winter is shorter and warmer, the Summer longer and cooler than at the North. The climate is admirable, allowing eleven hours of labor, daily, throughout the year, without injury. Its mildness relieves those employed of much expense for clothing, fuel and dwellings, unavoidable in more rigorous climates, while the excellence of the yarns and cloths produced, shows that the climate is favorable to the manufacture itself.

The fact that the increased consumption of the South in the past five years has been only 7,000, while the increased consumption of the North has been 35,400 bales, as stated by the eminent statistician Mr. Robert P. Porter, only proves that capital and experience are still at the North, not that the ultimate interests of cotton weaving will not be distributed mainly through the cotton belt.

It is believed by many that, if the protection to machinery intended for the manufacture of cotton was removed, and such machinery admitted free of duty, the lack of capital at the South would no longer be a barrier, and the ultimate effect would be to move much of the capital now devoted to cotton manufacturing at the East to the South. Let the mills come to the cotton.

No cotton mills in the East, or in Europe, can show such dividends as those declared by the mills in Georgia in the past decade. In almost any locality in the State cotton can be gathered from the field, ginned at the mill, and before night of the second day, be manufactured into yarns, or, at least, into cloths; the cost of baling, bagging and ties being saved. In addition, the cotton is presented to the preparatory machines in a loose flocculent state, far better suited to the manufacture than that which has endured for months the violent compression necessary for economical shipment. A bale of cotton weighing 500 pounds, will average in the mills of Europe about 400 pounds of goods, and the consumer will be charged with all commissions, storages, freights and insurance on 20 per cent. of the original weights, which is useless waste; while a bale of yarn shipped from a Southern mill would yield the same weight to the consumer without any waste whatever.

Our factories can be supplied with cotton at one cent per pound less than the New England factories. The two items of purchase of cotton and sale of fibres at home will give a profit of two cents per pound on the cotton thus consumed.

Transportation facilities are ample. The cost of water-power, steam-power, building materials, fuel and subsistence is less in the South than in any other part of the United States where cotton can be manufactured. While New England has ceased to compete with the South in the manufacture of the coarser yarns (No. 14 and under,) Southern mills sell heavy sheetings at 6¾c. per yard; shirtings at 5¾c.; 8-oz osnaburgs of fine finish at 9½c. and make money. Can any Northern mill do this?

Building materials can be purchased here at half their cost in New England. Lumber here is ten dollars per thousand; there from $30 to $35 per thousand. Bricks can be laid in the wall here for $10 per thousand; there they cost from $16 to $20. Machinery combining all the modern improvements can be selected. Recent improvements in picking, spinning, sizing and weaving machinery, will enable us to manufacture goods twelve and a half per cent. cheaper than old mills now running.

The material to be used, can be bought here as low as in any part of the world; and at a cost of ten dollars per bale less than to the Northern and English manufacturers.

From the published estimates of the Augusta mills for the six months ending June, 1875, running 717 looms, they made over 20 per cent. on $838,567.39, the cost of their factories, an average of $1,169.55 per loom. This is a large mill. Mr. Steadman, of Georgia, an experienced cotton manufacturer, estimates that a small mill with a capital of $100,000 can make, as net earnings per annum, 21¾ per cent. on the capital invested. The following history of the largest cotton mill in the State, the EAGLE AND PHŒNIX, of Columbus, Georgia, is a striking exemplification of the statements made above. It disproves completely the common inference that this latitude is unfavorable to the production of finer fabrics, "the truth being that, in all pioneer industries, the ruder products are made first, while the finer are forthcoming as the business becomes more fully developed."

Table showing the gradual advance of the consumption of cotton at Columbus:

1865–66	Rebuilding the mills.		1873–74	8,952	bales.
1866–67	80	bales.	1874–75	9,628	"
1867–68	566	"	1875–76	*12,118	"
*1868–69	2,207	"	1876–77	10,748	"
1869–70	1,927	"	1877–78	12,792	"
1870–71	4,953	"	1878–79	14,365	"
1871–72	6,830	"	1879–80	19,000	"
1872–73	7,428	"			

Mr. Wm. H. Young, the oldest and most successful manufacturer in the State, who is still at the head of the Eagle and Phœnix Co. estimates that the mills at Columbus have an advantage of one and nine-tenths cents per pound in the cost of raw material over their Northeastern competitors, and that, for a large mill of 1600 looms, this advantage amounts to over nine per cent. on the entire capital, or $120,099 per annum. This is $9.31½ per bale, estimating the annual consumption of this mill at 12,900 bales of 400 pounds each. These mills, like most of those in the South, were built when building cost most, and the capital of $1,250,000 represents the actual capital invested. Its earnings in 1877, a very unfavorable year, were 12 per cent. or $150,198.28 upon its entire capital, notwithstanding low prices for goods and the general business depression. This was done with 713 looms, 20,300 cotton spindles and 2,300 woolen spindles.

Its past operations, especially during the five " panic years," guarantee great success in the future, when, with the same capital, its production is doubled.

In regard to the class of goods manufactured by the Eagle and Phœnix Mills, running 45,000 spindles and 1,800 looms, and employing 1,800 operatives, and with an annual wages account of $400,000, they produce the better class of woolen and cotton fabrics which will

---

* Tallahassee (Ala.) Mills bought in this market.

vie with any, no matter where made.   Over one hundred varieties of
white and colored goods are turned out—checks, ginghams, calicoes
and others, whose names are known only to the dry goods trade.
Their cassimeres are elegant.   Besides, it is the only mill in the
United States which makes cotton blankets—white and colored

They use 13,000 bales of cotton per annum and 800 pounds of
washed wool daily.

The following table is taken from the annual statement of the Company on January 1st, 1880.

The total cash put in to date by stockholders is..$1,137,340 00
The total amount of dividends paid by the company from 1868 to date, is.................... 1,000,000 00

Assets.

The fixed investment is......................$1,897,418 44
The quick capital available in sixty days........ 1,306,566 67

Total..................................$3,203,985 11

Liabilities.

Debts, long and short maturities..$1,248,553 47
Machinery notes due in 1862..... 182,915 00—$1,431,468 47

Surplus as regards the public..............$1,772,516 64

The great success of the Sibley Manufacturing Company at Augusta,
the dividends 12 per cent., paid by the Langley Mills last year,
while the Enterprise Mill paid 9 per cent., and the superb manufactories at Columbus, form the strongest arguments for the erection of
new mills.   *The Legislature exempts all cotton manufactories from
taxation for a period of nine years.*   There are now 42 cotton mills
in Georgia.

THE AUGUSTA FACTORY.

Augusta is one of the oldest places in Georgia, having been laid
out by General Oglethorpe in 1735.   The population of Augusta was
fifteen thousand three hundred and eighty-nine in 1870, and increased

by 1873 to nineteen thousand eight hundred and ninety-one. According to a census taken for the City Directory of 1877, the population of Augusta is twenty-thousand seven hundred and sixty-eight. It is connected with Atlanta, Athens and Macon by rail; with Savannah both by rail and water; and by rail with Columbia, Charleston and Port Royal in South Carolina. Augusta receives about 200,000 bales of cotton annually. An Ice factory, manufactories of Commercial Fertilizers, numerous Cotton factories, Agricultural machine factories, etc., etc., attest the thrift and energy of the people. About 250,000 watermelons are shipped from this place to the North annually.

When the changes now contemplated are complete, Augusta will have 80,000 spindles, and Columbus nearly as many. Georgia and Massachusetts, which have been widely separated in politics, will in time become closely affiliated in sympathy as respects industry.

THE AUGUSTA FACTORIES.

On the first day of July, 1858, thirteen gentlemen bought of the City of Augusta, the property which had belonged to the Augusta Manufacturing Company, for $140,000, on a credit of ten years, bearing 7 per cent. interest, payable semi-annually, and one-tenth of the principal to be retired annually toward this enterprise; they contributed $60,000, which was expended the first two years in repairs and improvements. The interest was promptly met, and the payment of the principal was anticipated by five years from the profits of the company. The company invested largely in real estate and new machinery, and, by the surplus which it had accrued, increased its capital so as to give shareholders three shares for one, and make a working capital of $600,000. Since the war the Augusta factory has paid dividends amounting to 226 per cent. on its capital of $600,000, or $1,356,000 in cash, and the corporation now owns property worth $1,000,000. The factory has now 24,200 spindles and 800 looms making plain goods (sheetings and drills) from Nos. 12 and 14 yarns. The capital stock is $600,000. The gross earnings of this mill from June, 1877 to June, 1878, was $130,447.77, less expense on account of taxes, water-rates, interest, general expense, and insurance of $56,878. They paid four quarterly dividends of two per cent., amounting to $48,000, and carried to profit and loss account, $25,-469.77. The gross earnings amounted to 12 24-100 per cent. of the capital stock. The factory is five stories high, about four hundred and eighty-eight feet in length, and fifty-two feet in width. It has never paid less than eight per cent. The annual report for the last year shows the gross earnings for the year to have been $157,471, and the balance credited to profit and loss account was $318,198. The bonded debt was reduced $22,000. The capital stock is $600,000, and bonds are issued to the amount of $153,000. The total assets amount to $1,080,240, of which $850,418 includes mills, machinery and real estate. The goods manufactured during the year were: 6,393,284 yards 4-4 sheetings, 2,617,448 yards 4-8 sheetings, 2,608,680 yards 3-4 sheetings, 2,326,983 yards 36-in. drills, and 1,315,096 yards 37-in. drills; a total of 15,161,491 yards, weighing 4,727,591 lbs. The average number of looms running was 784, and the average number of hands employed 676. Four dividends were paid in the year, aggregating $120,000.

Consider the question of cotton, suggests an Augusta paper, as above presented in connection with the Augusta mill of 24,200 spindles and its $600,000 capital. In the year 1878 they consumed 11,819 bales of cotton, the average weight being 456 lbs. per bale. Upon the basis of an assumed saving of seven dollars per bale, this mill realized the sum of $82,733, or about 63½ per cent. of their gross earnings. If they had been deprived of this item of saving, they would have failed of the net profit of 12 24,100 per cent. and have incurred a loss of $9,264, which condition would compare favorably with some of our Northern mills during that memorable year.

The Langley Manufacturing Company was organized March 21st, 1870, with a capital of $300,000, which was paid in during the year, and put 6,400 spindles in operation. The following March the capital was increased to $400,000, and 3,200 spindles were added, the last payment being made December, 1871, so that the mill, with 9,600 spindles and 300 looms, got in operation in March, 1872, with its entire capital expended. The earnings for the next five years, say 1872 to 1876 inclusive, amounted to $318,833.64, less paid out for interest, $25,107.80, making the net earnings for five years, $293,725.84 Since 1876 there has been each year a small balance to the credit of interest, and the net earnings in 1877 were $37,215.48; net earnings in 1878, $45,000.64; net earnings in 1879, $81,277.31, making the net earnings for eight years, $457,218.27, on a capital of $400 000. Leaving out the item of interest which was incurred by the mill commencing business without commercial capital, and the real earnings for eight years were over $480,000, or an average of 15 per cent. a year on the capital, and the property is in better condition to day than when it commenced running, and now works 329 looms and 10,880 spindles. These figures are given as examples to show how well manufacturing pays in our section.

EXEMPTION FROM TAXATION.

In 1872, the Legislature of Georgia passed an Act to encourage the manufacture of cotton and woolen fabrics in the State of Georgia, by which Act it was declared that "Any mill or mills within said State, for the manufacture of fabrics out of cotton or wool, or both, whether such investment be applied in the establishment of a new factory or in the extension or enlargement of a now existing factory, shall be exempt from taxation for State, County and Municipal purposes, on the capital so invested, and on any property purchased or erected therewith, intended for and necessary to such manufacture, for the term of ten years from and after the laying of the foundation of the mills so to be erected."

COTTON MANUFACTURE IN GEORGIA AND NEW ENGLAND COMPARED.

(From the Augusta, Georgia, Constitutionalist).

Some time ago Gov. Shaw, of New Hampshire, published an elaborate argument in the Northern papers, to prove that the manufacture of cotton could be carried on much cheaper and better in New England than in the South. In an Augusta paper, we find, however, an absolute, positive and specific overthrow of his position in toto and detail. Such experienced manufacturers as Messrs Cogin, of the Augusta factory, Hickman, of the Graniteville, and Sibley, of the Langley, have been interviewed by an enterprising person on the *Chronicle*, who elicits the following pleasant information:

Reporter.—"You have doubtless noticed, Mr. Cogin, what has been said lately in regard to Southern and Northern manufactories? What is your experience in regard to the relative advantages of the North and South, for the location of cotton manufactories?

Mr. Cogin.—"There is no question but that the South possesses vastly superior advantages in many ways. We have one of the best climates in the world. The atmosphere has just the proper humidity for manufacturing purposes. Now, at the North, the air becomes so dry that steam has to be introduced into the weaving room to dampen the atmosphere, so as to prevent the threads from breaking. We never have any such trouble as that here.

"Again, the mills often have to stop because the water-courses are frozen up. This never happens at the South, and we can therefore run uninterruptedly. We can get plenty of excellent white labor;

in fact, it is much better than that which the Northern mills now have. It is equal to the 'Yankee' labor the Northern mills used to have, but which they don't get now. We can make more yards of cloth per loom than they can, running the same number of hours as they do, and we can, therefore, afford to sell it cheaper. Our water-power is plentiful, and cheaper on the average than at the North. They can't begin to compete with us while they have to use steam.

"It cost less than $6 per horse-power here for water, while at Fall River, where steam is used, the cost is $12. It wouldn't pay the Augusta factory, for instance, to use steam instead of water, if all the necessary fuel were put down at the factory free. The operatives in the Augusta factory work eleven hours a day. There is a superabundance of white labor here, and we never have had a machine stopped for the want of help during the nineteen years I have been with the Augusta factory. If we were to start a mill of the same size of ours to-day, we would have sufficient skilled labor in two weeks to run it."

In all the country, North or South, there has been no more successful enterprise than the Graniteville factory, under the management of Mr. H. H. Hickman, as President. What he said, therefore, must necessarily have much weight.

"In reference to cotton manufactories, Mr. Hickman on the subject said there could be no comparison between the North and the South. The South will eventually drive the North out of the market in brown goods, sheetings and shirtings. It is practically doing it now. The North is building no new mills for the manufacture of these goods. When Northern mills were compelled to sell their goods at cost, he could sell at a fair profit. He had no commissions to pay to agents to buy cotton, as Northern mills, because he bought it himself; more than half of it right at the mill. Getting the cotton right here, he had, of course, no freight to pay, as was the case with Northern mills, and he was satisfied that he could buy cotton to better advantage than the agents of those mills; in fact, he was assured that he made half a cent a pound in this way. He could get plenty of white labor, and cheaper than Northern mills could. His operatives could live for one-half the expense of those at the North. The latter used four times as much fuel, at twice the price per cord, while provisions were as cheap here as in Massachusetts. To sum up then: First, labor is cheaper; second, the operatives can live cheaper; third, he has no commissions to pay for buying cotton; fourth, he has no freight to pay on cotton; fifth, the larger proportion of goods are sold without paying commissions, and sixth, he could run his mill all the year. The Graniteville mill has not stopped two weeks in eleven years on account of water or weather. He finds sale for eighty per cent. of his products at home. He has sold sixty thousand dollars worth of goods to Knoxville, Tennessee, alone, in one year. He will build a new mill with the surplus of the Graniteville Company, without calling on the stockholders for a dollar, and he will be able to run it at three-fourths the expense in proportion to its size, that it cost to run Graniteville, because it is a modern mill with all modern improvements. He gets lumber at $6.00 per thousand feet, and bricks at $8.25 per thousand."

Mr. Sibley, President of the Langley Mills, says:

"That Africa and South America and the United States, have awarded the contract for sheeting needed for the Indian supplies, to the Langley Standard Sheeting made in this vicinity for some time. This contract was let out in New York, and the goods delivered there, thus competing successfully with New England." In regard to the labor, he says: "As to labor, I have been President of the Langley Manufacturing Company since 1870, and have had no difficulty in getting as good and reliable white labor as there is in New England, and who cheerfully work eleven hours per day, and could obtain more if we had any use for them; and many of them are Southern born, and have learned their trade in our own mill." He challenges any mill in New England to show as great a production of goods per loom and yarn per spindle, (on the same style of goods), or a cheaper cost of manufacturing. He concludes by saying that "the South has the best climate for manufacturing, the water power, the cotton, the men and women necessary to successful manufacturing. She lacks the capital, but, notwithstanding that, she has competed successfully with New England, in the manufacture of brown sheetings, shirtings and drills, both for the home and foreign trade."

NOTES ON COTTON MANUFACTURING.

Thirty-four factories in Georgia are exempt from taxation. They own property worth $4,138,875.

In Georgia there are to-day 213,157 cotton spindles in operation, and of them Columbus claims 60,000. The Eagle and Phœnix Mills of that city alone operate 44,000, about 20,000 more than are operated by any other one factory in the State.

Athens has in its immediate vicinity 2,994 horse powers in the streams near that city, of which only 395 horse powers are utilized. The Georgia Factory uses the largest amount, 125 horse powers, and has a reserve of 275.

Cotton goods manufactured in the South, are now selling in New England. Manufactories of cotton goods are springing up where new mills are exempt from taxation for ten years. Houses are built by the mill owners, and free rent and the use of a garden plot are the conditions of service.

Factory stocks have all advanced at Augusta. Graniteville is now quoted 145 bid, 150 asked, and Augusta and Langley 148 bid, 150 asked.

Four hundred workmen are at work on the new Sibley Mills at Augusta, Georgia. The pay roll amounts to $2,000 a week. It will have 24,000 spindles and will disburse annually among its operatives $175,000. It will use about eight thousand bales of cotton yearly, and yield a product of not less than $1,000,000, or over $3,000 daily. The building of such a mill will add several thousand persons to population from the industrial walks of life, whose various wants must be supplied here. The price of every horse-power supplied by the canal, has been fixed at the reasonable sum of $5.50 per annum.

The stockholders of the Graniteville (Ga.,) Factory held their annual meeting Thursday last. The Vaucluse Factory has been built without a dollar's cost to the stockholders. It is one of the finest factories in the South for fine class fabrics. It has 10,000 spindles, and cost $347,000. The statement of the operations of the year of the Graniteville Mills, show that the profits are $127,774.67. The Vaucluse Mills profits are $66,800.16. The total profits are $194,574.83. The expenditures of both mills were $51,045.82. The profits, less expenditures, equal the total net profits. Dividends paid 18,000. Carry to profits and loss the balance making net profits, the amount credited profit and loss, $125,649.76. The mills produced 16,164,079 yards, consuming 13,000 bales.

It has been suggested that it will be well for cotton mills to turn their attention to the manufacture of cotton canvas. Not a yard of this canvas is now made South of Baltimore, and vessels trading with Southern ports would buy from Southern Mills.

NOTE.—The Comptroller-General's report for 1880 is liable to mislead the uninformed. This report states that cotton manufacturing has decreased in one year $1,132,673. The Columbus factories reported to the city council for taxation their sales for 1878 at $891,135, and for 1879 at $1,397,722. This increase of half a million does not include the sales of the 4,000 spindle mill located three miles above the city. A similar result is shown in Augusta, where there are 44,000 spindles. Columbus has 60,000 spindles. The apparent decrease is due to a change in the form of the returns. Previous to 1876 the assessors obtained the capital stock of mills and reported them ; also business. Now the law requires them to report real estate, machinery, notes, accounts, and other data, the same as other business. "Cotton manufactures" embraces only the returns of machinery, which is taxable, and the State Constitution exempts both new mills and machinery from taxation. Much that comes under the caption of merchandise, personal property, and other entries of 1879, were credited to "cotton manufactures" in 1878.

THE AUGUSTA CANAL.

The following description of the Augusta Canal is taken from Derry's Georgia :

"Its dimensions and capacity are as follows: Length of main canal, or *first level*, seven miles ; and including *second* and *third levels*, nine miles. Minimum water-way, one hundred and fifty feet at surface, one hundred and six feet at bottom, and eleven feet deep, making an area of cross-section of fourteen hundred and eight square feet. The bulkhead, locks, dam, and other structures are composed of stone-masonry formed of granite rock laid up in hydraulic cement mortar, and are of the most substantial character. The area of openings for the supply of the canal amounts to fourteen hundred and sixty-three square feet, and the entire waters of the Savannah River are made available for maintaining the supply. There are about two hundred and seventy-five acres of reservoirs, exclusive of the canal proper and the pond above the bulkhead dam. There is a bottom grade or de-

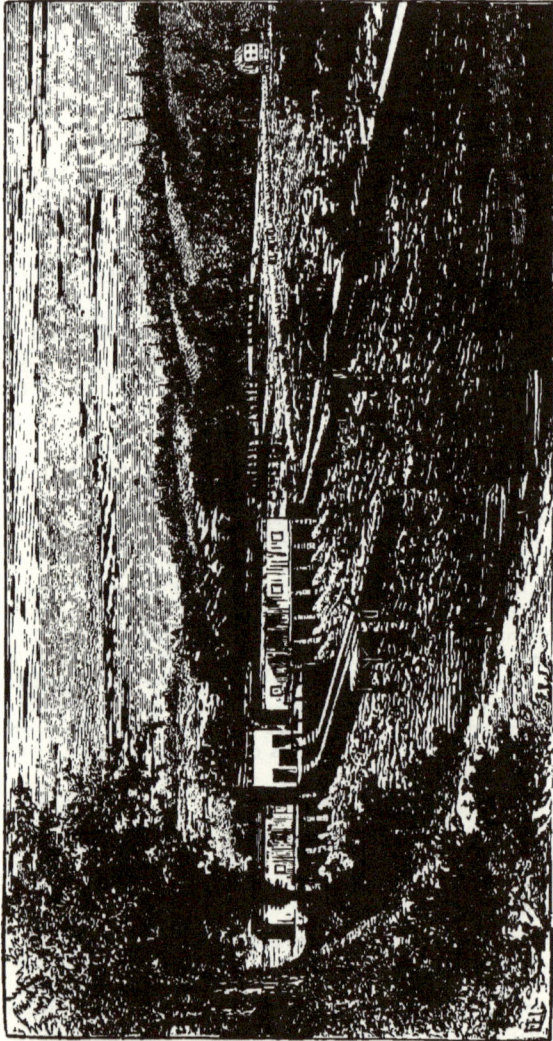

BULKHEAD, LOCKS, AND DAM OF AUGUSTA CANAL.

scent in the main canal of one hundredth of a foot in one hundred feet, giving a theoretical mean velocity of $2\frac{74}{100}$ feet per second, or a mechanical effect under the minimum fall, between the *first* and *third levels*, or between the first level and the Savannah River, below Rae's Creek, of upwards of fourteen thousand horse-powers, not including available supply from the surface of the reservoirs. Of this immense power but nineteen hundred horse-powers are contracted for, leaving at least twelve thousand horse-powers to be disposed of."

FOUNTAIN IN FORSYTH PARK, SAVANNAH.

Savannah has been appropriately called the "Forest City," The city is filled with little parks, each ornamented with a fountain, a monument or mound. In one is the marble obelisk to General Nathaniel Greene; in another, the memory of brave Sergeant Jasper of Revolutionary fame is perpetuated; while in Monterey Square stands the elegant monument in honor of Count Pulaski who was killed while fighting for American liberty, on the 9th of October, 1779. The shade trees are chiefly the growth of the forest. The fountain in Forsyth Park is modelled after one in the Place de là Concorde, in Paris.

The walks are prettily arranged and covered with shell. Bonaventure Cemetry, three miles from the city, is one of the loveliest spots in the country; long avenues arched by the branches of great live oak trees from which an immense quantity of gray moss sweeps, adding much to the solemnity of the place. Bonaventure derives its name from the original tract of which it formed a part, and which was settled about 1670 by Colonel John Mulrayne.

The streets of Savannah are broad, shaded by great water-oaks and other indigenous trees that are green the year round.

This city is situated on the river of the same name, eighteen miles from the sea, with a capacious and well protected harbor, with from seventeen to twenty-one feet of water at high and low tide. Improvements are now being made in the river with a view to obtaining depth sufficient for any vessel.

Savannah has a population of from 30,000 to 32,000 inhabitants. It is the second largest cotton port in the United States, while its shipments of rice, lumber and naval stores are immense.

The city has ample transportation facilities; the Savannah and Charleston Railroad connecting Charleston and the North; the Central (Georgia,) to Augusta, Atlanta and the Northwest, while the Savannah, Florida and Western Railway connects with the growing sections of South Georgia and the whole State of Florida. The fine steamships of the Ocean Steamship Company make semi-weekly trips to New York, while Philadelphia, Baltimore and Boston are connected by weekly lines of steamships of great capacity and elegant accommodations.

Especial attention has been given to its sanitary condition.

BRUNSWICK.

Brunswick, the terminus of the Macon and Brunswick Railroad, is one of the oldest towns in the State. The harbor is equal to any on the Atlantic coast, and it is the hope and belief of the citizens that the port will at no distant day assume the prominence to which it is justly entitled.

THE SEA ISLANDS.

The SEA ISLANDS of Georgia yield the celebrated "sea island cotton" which is worth much more in the markets than the upland or "long staple" cotton. Cumberland and St. Simon's Islands are especially noted.

Olives are successfully cultivated, and oil made from the olives grown on St. Simon's Island has been pronounced by competent judges not inferior to the best productions of France or Spain. The oil crop from these islands is annually sold at from $6 to $8 per gallon.

The scuppernong grave thrives well. The soil and climate of the islands are peculiarly adapted to its perfect development.

TIMBER AND VALUABLE WOODS.

NORTHERN AND MIDDLE GEORGIA.

The walnut, cedar and white oak timber of Northern and Middle Georgia is exceedingly valuable.   Cedar 4x4x12 feet long is worth in the Baltimore market $1.25 a cubic foot, or 20c. per foot.   Cedar can be handled as yet in Baltimore only in a small way, but the New York and Philadelphia markets offer better inducements to the trade.   Sale for walnut logs 20 inches or more in diameter, and 12, 14, or 16 feet long, is readily found in the Baltimore market at prices ranging from $50 to $60 per thousand feet.   Quantities of walnut timber obtained in Southwest Virginia, East Tennessee and Kentucky, are being shipped to England by agents employed by British Cabinet makers.

The time is not far distant, it is to be hoped, when we will cease to burn these valuable woods, in order to make new "clearings" for field crops.   We should cease also, in a measure, to ship timber suitable for furniture, wagons and farm utensils to the North, and ship them back in the form of manufactured articles in daily use, thus paying freights twice and commissions four times.   The cost of transporting pine lumber to Baltimore is $7.50 per thousand feet.

The Geological Museum at Atlanta contains specimens of 130 varieties of indigenous woods, and there are upwards of 100 others that are not embraced in the collection.   The bottom lands all over the State are well supplied with white oak and kindred woods.

A rich reward awaits the manufacturer of cheap furniture and household utensils in Georgia.   Beds, tables, buckets, broom-handles, axe-helves, wheel-spokes, etc., etc., may be sold to the laboring classes with great profit to the manufacturer and saving to the laborer. It is very probable that skilled and experienced workmen, who will stand at the lathe and work, themselves, can easily find landholders who will invest part of the capital required, and sell on time as much land as may be wanted, in order that the mechanic so employed may have his home and add to its value whenever the opportunity offers. A land owner having 500 acres well timbered, can thus afford to sell 100 acres at $5 per acre, rather than at $10 per acre to the ordinary unskilled laborer.

Iowa walnut logs are now being shipped to England.   The timber is bought in the tree by a Liverpool agent, and is cut down and hewn square.   The ends are then painted to prevent cracking from exposure to the weather.   The logs are shipped to Liverpool to supply the cabinet makers of Great Britain.

Of the great demand in England for black walnut timber for furniture purposes, the Des Moines (Iowa) *State Register* says:

"The lumber dealers of England are making a grand raid on Iowa and all the black walnut States, and are fast taking from us all that we have left of that timber. The havoc of timber in Ohio and Indiana—the settlers there spending nearly a hundred years in steadily destroying the woods with which they found the surface covered, girdling and killing the great forests one year and burning them the next—is one of the queer things in American history as it is now looked at."

The above is very suggestive to the people of Georgia.

LUMBER.

(From the Baltimore Market Journal, April 3d, 1880.)

These quotations represent prices at which lumber can be sold by the cargo.

Yellow Pine.

Virginia and Maryland, 3-4 box	$ 8.00 @ 10.00
5-8 " "	6.00 @ 7.00
4-4 Virginia flooring, dry and clear	14.00 @ 15.00
4-4 " rough, best	11.00 @ 12.00
4-4 " stained, not dry	9.00 @ 10.00
Small Joists, Virginia	10.00 @ 11.00
Large " "	11.00 @ 12.00
3x4 Scantling	10.00 @ 11.00
Georgia and Florida timbers, re-sawed, bills cut to order	22.00 @ 25.00
River Flooring	12,00 @ 15.00
Southern Siding and Edge Boards	14.00 @ 16.00

White Pine.

Michigan and Wisconsin—1st, 2d and 3d clear, 4-4	$45.00 @ 47.00
1st, 2d and 3d—5-4, 6-4 and 8-4	46.00 @ 48.00
Edge Culling	17.00 @ 18.00
Stock Culling	18.50 @ 19.00
White Pine, Susquehanna	
Selects and better	42.00 @ 47.00
Selects and Picks	35.00 @ 40.00
4-4 Boards, good run of log	22.50 @ 25.00
4-4 Stock Barn, 12 inches	21.00 @
4-4 Edge Samples	16.50 @ 17.00
4-4 Stock do. 12 inch	17.00 @ 18.00
5-4, 6-4 and 8-4 run of log	22.00 @ 24.00
4-4 Edge Cullings	18.00 @ 19.00
4-4 Stock do. 12 inch	19.00 @ 21,00
Hemlock Fencing, 16 feet, 1x6	13.00 @
2x3, 2x4, 16 feet	13.00 @

Walnut.

Walnut, Indiana, ¾ wide, 7 or 14 feet, 1st and good 2d, Coffin Stuff	$65.00 @ 70.00
do. do. ¾ wide, 1st and good 2d Cabinet Stuff	50.00 @ 62.50
4-4 wide, strictly 1st and 2ds, thoroughly dry	70.00 @ 75.00
5-4 and upwards	70.00 @ 80.00
Scantling, 4-4 and 4-5	70.00 @ 75.00
Newel Stuff, all prime, 8 to 16 feet long, 6-6, 7-7, 8-8, 9-0 and 10-10, dry 1st and good 2ds	75.00 @ 85.00
Culls, half-price	

Cypress Lumber.

Boards, 1x6-16 feet, clear	$18.00 @ 20.00
" 1x6-16 feet, fencing	12.00 @ 12.50
" 1x12 to 18 inches, 16 feet clear	18.00 @ 22.00
Boards, rough, wide	10.00 @ 12.00

Poplar.

Poplar, ¾ West Virginia	$16.00 @ 19.00
" 4-4	22.00 @ 25.00
" Thick	25.00 @ 30.00
" ⅝ Indiana	22.00 @ 25.00
" 4-4 "	27.00 @ 30.00
and upwards	30.00 @ 33.50

Cherry.

1 to 5 inches thick, as wide as possible......$40.00 @ 50.00

Ash.

Ohio and Indiana, tough white, 1 to 5 in.
thick, wide, and, if possible, 15 feet
long, (culls half price)......................$32.50 @ 40.00
Pennsylvania and West Virginia............. 30.00 @ 35.00
North Carolina, 1 to 5 inch Plank.......... 18.00 @ 22.00
Balusters, N. C., 2 and 2½ inches............ 20.00 @ 25.00

Hickory.

Good White and Tough Hickory, 1 to 3
inches thick............................. 40.00 @ 50.00

Cypress Shingles.

Hearts, strictly No. 1	$6.50 @ 7.00
" seconds good	4.75 @ 6.00
" Saps as in quality	3.50 @ 5.00

Barrel Staves.

White Oak, heavy, 34 in. long, 4½ to 6 in.
wide, 1½ in. to 2 in. on heart............   20 @   23
do. light, ¾ in. long, 4½ to 6 in wide, 1 to 1½
in. on heart........................... 16.00 @ 18.00
Red Oak, heavy, 35 in. long, 4½ to 6 in. wide,
1½ to 1½ in. on heart.................. 12.00 @ 15.00
do. light, 33 in. long, 4½ to 6 in. wide, ¾ to 1
in. on heart...........................   9.00 @ 12.00

Western White Oak Staves.

Barrel, heavy, 34 in. long, 4 to 6 in. wide, 1½
in. on heart........................ 24.00 @
do. light, 34 in. long, 4 to 6 in. wide, 1 in.
thick, on heart..................... 20.00 @ 22.00
Heading, 22 in. long, 8 to 13 in. wide. 1½ in.
on heart .... 20.00 @ 25.00
Pipe, light, 54 in. long, 4½ to 6 in wide, 1¼ to
1½ in. on heart.................... 60.00 @ 75.00
do. heavy, 56 in. long, 4½ to 6 in. wide, 1 in.
on heart........................... 85.00 @100.00
do. extra heavy, 60 in. long, 4½ to 6 in. wide,
2 in. heart .... 125.00 @150.00
do. extra heavy, 66 to 72 in. long, 5 to 8 in.
wide, 2 in. on heart............... 175.00 @225.00

Bucked Staves.

White Oak, 34 to 35 in. long, ⅜ in..........$15.00 @ 16.00
"   "   "   "   " ¼ in.......... 18.00 @ 21.00
Machine Flour Barrel.................... 7.50 @ 8.50
Heading, soft and hard.................... 4.50 @ 6.00

Laths.

White Pine 4 ft. Laths, dry............$ 2.20 @ 2.25
Spruce   "   .... 1.90 @ 2.00

Wood Hoops.

Hogsheads, shaved, 14 ft. long, 1¼ to 1½ in.
   at small end......................$18.00 @ 20.00
Tight Barrel, 8 to 8½ ft. long, 1 in. at small
   end...................... 8.00 @ 9.00
Flour Barrel, hickory, 6½ to 7 ft. long, ¼ in.
   at small end...................... 3.50 @ 4.50
Poles, Hogsheads, 14 ft. long, 1¼ to 2 in. at
   small end .... 30.00 @ 40.00
Poles, Tight Barrels, 7½ to 8 ft. long, 1 to 1½
   in. at small end...................... 12.00 @ 15.00

FREIGHTS.

(From the Savannah (Ga.) News, March 12, 1880.)

LUMBER—*By Sail.*—To Baltimore and Chesapeake ports, $5.50@6 00; to Philadelphia, $6.00@6.50; New York and Sound ports, $6.50@7.00; to Boston and eastward, $6.50 @7.60; to St. John, N. B., $8.00; [Timber from $1.00 to 1.50 higher than lumber rates]; to the West Indies and windward, $7.00@8.00; to South America, $17.00; to Spanish ports, $14.00@15 to United Kingdom, for orders, timber 33@34s., lumber £5 5s@£5 10s. From 50c. to $1.00 additional is paid here for change of loading port.

NAVAL STORES—*Sail.*—Rosin and spirits, 3s. 3d.@5s. to United Kingdom or Continent; to New York. 35@40c. on rosin, 60c. on spirits. *Steam.*—To New York, rosin, 30c., spirits 80c.; to Philadelphia, rosin, 40c., spirits. 80c.; to Boston, rosin 40c., spirits, 90c.

TIMBER AND NAVAL STORES.

SOUTHERN GEORGIA.

It is stated that within 60 to 70 years past the forests of the Eastern States have disappeared, and that they have been consumed by a population less dense than that which is draining the present supply. The census report of 1870 stated the annual production of lumber to be 12,000,000,000 feet.. A standard statistical work then estimated the country's total supply of standing pine, spruce and hemlock timber, suitable for lumber, at 100,000,000,000 feet. It is probable, if the present rate of consumption continues, that all of the territory of Canada, and of the United States, east of the Rocky Mountains, will be practically denuded of merchantable timber, except that in the Southern, Atlantic and Gulf States. Georgia, from its geographical situation and transportation facilities, offers better inducements to investors than any of the Atlantic or Gulf States.   The price of lumber in Wisconsin, April 17, 1880, was $4.25 more per thousand feet than at same date in 1879, which indicates a rapidly increasing demand. Michigan saws more than 3,000,000,000 feet per annum.   The pineries of the Northwest are already beginning to show signs of exhaustion; the annual consumption now being estimated at 13,000,000,000 feet. Inferior white pine lands, bought seven years ago in Pennsylvania, at $2.50 per acre, have been sold this year for $18 to $19 per acre. Michigan lands have also advanced several hundred per cent.   Pennsylvania is almost denuded of exportable timber.   "Texas, New Mexico, Arizona, Colorado, Kansas, Nebraska, Dakota, Eastern Montana, Illinois, Iowa, the west half of Missouri, Minnesota west of the Mississippi, and Southern Wisconsin, *are chiefly prairie lands and are almost treeless territory.*   This vast Western country, with Ohio and Indiana, is largely dependent on the small amount of pine in Michigan, Northern Wisconsin, and Northeastern Minnesota.   Ten years will probably exhaust all that Michigan can supply."

The shipment and sales from the lumber districts of the United States, this year, are greatly in excess of any previous year since 1872.   There are more than 20,000 square miles of forest land, containing exportable timber, in Southern Georgia.   Lands which would sell, if they were located in Canada, Michigan, or Pennsylvania, at $10 to $15 per acre, I am authorized to offer at from $1.50 to $3.00 per acre.   They are distant only 150 miles from the coast, and raftable streams abound.   The South is the only part of the Union of which it can be said pine lands are as cheap as they were fifty years ago.   Immigration and the foreign demand will soon change the present low prices, and bring thrift and wealth to present landholders.

These facts are given in order to call the immediate attention of turpentine men, capitalists, and lumber-men to the great section through which the Macon and Brunswick, and Savannah, Florida and Western Railroads run.   Thousands upon thousands of acres, within easy distance of these great avenues, can now be bought at prices which will yield many hundred per cent. in the near future.

From information derived from reliable sources, we estimate the amount of timber and lumber marketed in Georgia in the year 1880, to be 300,000,000 feet, worth probably $5,000,000 or more.   The following figures are from nine representative lumber merchants in Georgia, and are taken collectively, in order to suggest the present status of this important industry:

Total number of feet cut per day, 338,000; total number of feet cut per year, 65,000,000; total number of acres exhausted per year, 60 000; total number of hands employed 1,135.

It is estimated that the annual product of Georgia in the rosin and turpentine trade is 200,000 barrels of turpentine and 300,000 barrels of rosin, which would give to the State of Georgia alone for this one industry $3,000,000—nearly half as much as was produced by the whole country in 1876. *The Albany (Ga.) News* makes the following interesting statement: Below we give a few figures obtained from twelve representative turpentine men in Georgia—the total amount of capacity, yield, acres under cultivation, the number of distilleries and stills, with the number of hands employed, this recapitulation being only submitted in order to give an insight into this great industry in our midst: Total number of crops, 325; number of acres, 76,000; number of boxes, 3,340,000; yield per year, $459,000; hands employed, 820; distilleries, 20; stills, 17.

TIMBER AND LUMBER EXPORTS, 1877.

Savannah	51,281,972
Darien	74,106,152
St. Mary's	18,116,000
Brunswick	19,092,410
Total	162,596,534

During the year 1877 there were 75,000,000 feet of timber down; in 1878 there were 54,000,000; in 1879 there were 50,500,000. Since the 1st of January of this year there have been 40,000,000 feet in market, divided up as follows: January, 5,066,000; February, 13,644,473; March, 7,356,713; April, 11,214,000; May, up to date, 2,700,000. The total value of the timber received so far this year, amounts to about $580,000; total value of timber arrived since January 1st, 1877, about $2,634,000.

The following are the quotations of the timber market, published in the Darien *Gazette:* Square, 600 average, $7.50@8.00; 700 average, $8.00@9.50; 800 average, $9.50@10.50; 900 average, $10.50@11.50; 1,000 average, $11 00@12.00; 1,100 average, $12.00@13.50; 1,200 average, $12.00@14.00. Scab, 300 average, $7.00@8.00; 400 average, $8.00@9.00; 500 average, $9.00@10 50; 600 average, $10.00@11.00.

Lumber men are rapidly buying up all the available land lying next to the railroad, and land is appreciating in value. They consume only the large trees and then offer the land on very favorable terms to settlers. Especial inducements will be offered to colonies, for it is best that immigrants shall have congenial society of their own selection from the time that they settle in a strange land. They will be cordially welcomed, and will soon cease to feel as strangers.

The Georgia Land and Lumber Company has invested $80,000 in the finest saw-mill on the coast, at St. Simon's Island, the timber supply for which must come from these lands. Settlers, while preparing homes and farms, probably can obtain work.

The rivers continue in fine rafting condition except for a time during the Summer. At Darien, during the months of June, July, August, and sometimes in September, the river is quite low, and but little timber is floated down. At no time in the Summer is the river so low that a few rafts cannot come down.

The following letters are here inserted, in order to give a description of the lands in this part of Georgia. The Georgia Land and Lumber Co. own 300,000 acres, and, hence, a description of their property is applicable to the whole section of country comprised in the "Wire Grass" counties of Georgia, except that lying near swamps.

Mr. Hebard, an expert of fifteen years' experience, now managing the largest saw-mill in the United States (1870,) after visiting the property, says:

"I never saw land better adapted for lumbering, or more tempting for settlement. It rolls as prettily as the prairies of Illinois without their unhealthiness, and the whole of it is accessible at all times of the year for the log-wagons and mules used in getting out the timber. For many years you can contract for the delivery on the cars of shipping and mill timber at from $2.50 to $4.00 per 1,000 feet, and with your anticipated freight contract with the railroad company, can deliver at Brunswick, alongside vessel, for $5 additional, say $8 to $10 on board. Timber can from thence be delivered in New York and Philadelphia for $8.

This is confirmed by Mr. Leighton, a veteran explorer, who has explored more land in Maine, Michigan, Georgia and Canada, than any other man in the United States, whose report is herewith annexed.

"Saginaw, Mich., October, 1869.

"A. G. P. DODGE, Esq.,

"Pres't Georgia Land and Lumber Co.

"DEAR SIR:

"These lands, situated in the counties of Laurens, Pulaski, Telfair and Montgomery, in the State of Georgia, are centrally distant about 125 miles from the port of Darien, on the sea-coast, are from 250 to to 500 feet above tide water, and lie between the Oconee and Ocmulgee rivers, both of which are navigable for steamboats above the whole property. The method of examination adopted by me was to visit each of the lots owned by the Company, to select an average acre from the lot, to measure all the trees upon it large enough to cut, and thus determine, with as much accuracy as possible, the quantity of lumber that could be manufactured from each lot. I also carefully noted the direction and volume of the streams, and the distance that logs would have to be hauled to them. The extreme distance to navigable or driving points does not exceed 2½ to 3 miles, while the average haul is less than 1½ miles. The expense of opening the streams not cleared I estimate to be from $50 to $100 per mile.

"This timber may be classed as follows: say 6,000 feet shipping and sawing timber, and the balance, 4,000 feet, as suitable for railroad ties, small spars, etc. The quality of the timber is unusually good, being sound, straight, and fine grained, as trees cut upon the land indicated, while some are unusually large; I have frequently found trees of 3, 4, and even 5 feet in diameter, and from 70 to 80 feet high to the first branches. The timber is of the valuable kind known as the long-leaf yellow pine. There being no other growth than the pine, the forest is open, and the ground being dry, no expense is necessary to open roads for hauling purposes. These lands are intersected by seven creeks, all of which, with some of their tributaries, are large enough to be available for all purposes of running or driving logs and long timber into the Altamaha and to tide water. The Macon & Brunswick Railroad runs through and near the center of these lands, for a distance of 40 miles, thus giving additional facilities for getting their products to market. The ground is covered with a natural grass, upon which cattle thrive throughout the year without other food. The water is good and the climate very desirable.

"In conclusion, permit me to say that—the timber standing being unusually large in quantity and fine in quality, the streams large enough for driving purposes, the haul shorter than common, and no expense necessary for opening roads—the facilities for getting out the timber upon these lands are unusual, and much superior to those of any other tract of land that has ever come under my observation.

"Very respectfully, your obedient servant,

"JOHN LEIGHTON."

"New York, March 23, 1870.

"Mr. John Leighton has, during the past twenty years, been frequently engaged by us in the examination and estimating of pine timber lands, and is now, at this writing, examining a large tract for us.

"We have always placed great confidence in his reports and estimates, and have made extensive purchases predicted on them, and have never had occasion to regret acting on his reports. In most cases the result has been that he has rather under than over estimated. He has been for over thirty years engaged in selecting, locating, surveying and estimating timber lands, and few men in the United States have had more experience or commanded more entire confidence from his employer than Mr. Leighton.

"Very respectfully

"PHELPS, DODGE & CO."

It is safe to say that it will not cost more to put this land in a condition ready for crops than the prairie land of the West, while there may be left standing a sufficient amount of timber to supply all the needs of the farm for a life time. In this connection we will explain the mode of "clearing land" in vogue in Georgia, as many may not understand how land can be cultivated unless all the trees are cut down and taken away. In Southern Georgia where no "undergrowth" exist, the large pines are killed with the axe by cutting around the trunk until a complete circle is made through the sap. The time required for this is, perhaps, five minutes to each tree, and the tree dies gradually; so that, after one season, the growth of the plants cultivated is not affected by the dead tree.

The "tap root" of the pine is very deep, the lateral roots being small in comparison to trees that, like the oak, are sustained chiefly by the lateral roots. The larger roots near the surface are cut, and, with the undergrowth, (where undergrowth exists, as is the case in every other part of the State,) are burned. The ashes fertilize the soil.

This wasteful destruction of humus—for the grass is also burned before the land is cultivated—should not be permitted except in cases of necessity.

Lumber and the material for building, are cheaper here than in any part of the United States, as the following prices will show.

First class "dressed" lumber, about $10.00 per thousand feet.
Second class   "   "   "   5.00   "   "
Brick for Chimneys, etc., etc.____$6.00 to $8.00 per thousand.
Cost of "framing" a house._____$8.00 per thousand feet.
    "   "   "flooring"   "   _____ 1.00 " square.
    "   "   "weatherboarding" a house.. 1.00 "   "
    "   "   "shingling" a house._____ 1.50 "   "

A cottage of three or four rooms of good size, will cost from $100 to $200   It is advised that each colony of immigrants include a carpenter and build their own houses. The houses will be built for them if they will send plans and specifications, when estimates of cost will be furnished them. The immigrant will be required to pay cash for such buildings, but they can easily build their own dwellings. They are offered a title *in fee simple* to land at a less price than tenants in Europe are required to pay annually for land that is no better.

The Savannah, Florida and Western Railroad, and the Macon and Brunswick Railroad, have developed this timber region, and the accompanying diagram will show the most available territory for this business:

The Savannah, Florida & Western Railroad is the shortest and most direct route to Southern and Southwestern Georgia. Its main line extends from Savannah, Ga., to Bainbridge, Ga., on the Flint River, two hundred and thirty-seven miles. The Albany division extends from Thomasville, Ga., to Albany, Ga., fifty-eight miles. The Florida division extends from DuPont, Ga., to Live Oak, Florida, forty-eight miles; making a total of three hundred and forty-three miles under its management.

CONNECTIONS.

Through cars from Savannah to Bainbridge, connecting with steamers for all points on the Flint, Chattahoochee and Apalachicola rivers.

Through cars from Savannah to Albany, connecting with through trains on Southwestern division Central Railroad of Georgia to Macon, Atlanta, Eufaula, and via Montgomery and Eufaula Railroad to Montgomery, Ala., New Orleans and Louisville.

Through trains via Main Line and Florida Division via Live Oak, connecting with trains on Jacksonville, Pensacola and Mobile Railroad and Florida Central Railroad for Jacksonville, Fla., without change; connecting at Jacksonville with steamers for St. Augustine, Palatka, Sanford, Mellonville, Enterprise, and all points on the St. Johns and Oclawha rivers; at Baldwin with trains on the Atlantic, Gulf and West India Company's Railway for Fernandina, Gainesville, Cedar Keys, Tampa, Manatee, Key West, etc.; at Live Oak with Jacksonville, Pensacola and Mobile Railroad for Madison, Monticello, Tallahassee and all points in Middle Florida.

Elegant Pullman Sleeping Cars daily between Savannah and Thomasville, Albany, etc.

Elegant Steamships ply regularly between Savannah and New York, Savannah and Boston, Savannah and Philadelphia, and Savannah and Baltimore, making quick and prompt connection with the railroads that termniate at Savannah. In Through Freights their facilities are unsurpassed.

Local freight and passenger rates in Georgia are governed by the State Railroad Commission.

Shipments of Naval Stores for the past eight years from line of Savannah, Florida and Western Railway, have been:

	1872.	1873.	1874.	1875.	1876.	1877.	1878.	1879.
Rosin, bbls.....	1,416	15,464	23,500	39,819	65,854	95,592	126.961	129,970
Sp'ts Turp'ntine	354	3,649	5,200	9,130	13,974	20,452	26,517	24,751
Total.....	1,770	18,113	28,700	48,949	79,828	116,044	153,478	154,721

The Macon & Brunswick Railroad was completed in 1879, and forms the shortest main line from the West and Southwest to the coast of Georgia. It is predicted that it will be one of the most important and prosperous railroads in the South.

The Macon & Brunswick Railroad possesses unsurpassed facilities as a Through Passenger Route from Chicago, Cincinnati, St. Louis, Louisville and other Western cities to this section of the South and Florida. Through tickets are on sale at all principal ticket offices, and Through Pullman Palace Cars are run both from Chicago and Cincinnati, via Nashville, Chattanooga, Atlanta, Macon and Jessup, to Jacksonville, Florida.

In Through Freights. they have the low rates of the "Green Line," a well known association, and a first-class steamship line from Brunswick to New York. There are also good facilities for shipping by regular sailing packets from Brunswick. These roads are also parties in rates, to what is known as the "Florida Dispatch Line." This is a very fast freight line, mostly used for vegetables and fruits, which are carried in ventilated cars built specially for the purpose, and the cars of which go by passenger trains, or on passenger train schedules. The rates of freight by same are exceedingly low ; in many instances merely nominal. It reaches to Chicago, Cincinnati, St. Louis, Indianapolis, Louisville, Nashville and other Western cities; and by Atlanta, Dalton and Charlotte to other large cities. The Southern Express also has its line over this route. The road, further, has good and full shipping facilities with Savannah, via Jessup.

METALS AND MINERALS OF GEORGIA.

(From the Report of Dr. Geo. Little, State Geologist.)

COPPER. — This metal exists in large quantities in the counties of Fannin, Towns, Cherokee, Paulding, Haralson, Carroll, Green and Fulton. Fannin county mines are an extension of the celebrated Ducktown veins in the State of Tennessee, and are considered equally rich in that metal.

IRON.—Iron ores, either hematite, limonite, or fossiliferous, are abundant throughout the Northern part of the State, and are found to a considerable extent in the counties of Harris and Talbot, in Middle Georgia, and in Burke of the Southern division. In the counties of Dade, Walker, and Chattooga, it lies side by side with large deposits of coal, thus affording ample facilities for smelting. In Bartow County we find the best brown hematite, which, in combination with manganese, also abundant in that section, forms that beautiful, mirror-like iron, called by the Germans *Spiegeleism.* The brown hematite is also abundant in Polk County. At the date of the last report, there were in the State 20 iron foundries, with a producing capacity of 300 tons per day, or 100,000 tons of pig-iron per annum, worth at the market price of $20 per ton, $2,000,000.

COAL — The discoveries of this mineral have been confined to the three counties of the extreme Northwest, to wit : Dade, Walker and Chattooga. The supply, though, is so abundant and accessible that it bids fair to be permanent, both for fuel and for mechanical uses. One of these mines yield 300 tons per day. Railroads, with connections leading to all points, have been constructed to the mouths of these mines. A superior article of coke is also prepared on the spot, and shipped to the smelting furnaces of this State and Tennessee.

SLATE, admirably adapted to roofing, exists in large quantities and at points accessible. At Rockmart, Polk County, immense shipments are made annually to the various cities and towns of Georgia and the adjoining State. A railroad connection with the main trunk runs directly to the quarries.

LIMESTONE.—Immense beds exist throughout all the Northwestern counties, and there is a fair distribution of it in nearly every section of the State. At Kingston, Bartow County, Mr. G. H. Waring is largely engaged in the manufacture of Hydraulic Cement, an article that has come into extensive use, and has won a deserved popularity.

MARBLE.—This mineral exists in exhaustless quantities and of many varieties. It varies in quality from the fine statuary to the coarse-grained used for building. The black marble is found at Tunnel Hill, on the Western and Atlantic Railroad; the red at Dalton; the pink at Varnell's Station, on the East Tennessee and Georgia Railroad, and in Whitfield County. The white, of best quality and in immense supply, near Jasper, Pickens County, on the line of the Marietta and North Georgia Railroad, now in course of construction ; also at Buchanan, in Haralson County, and at Van Wert, Polk County.

SERPENTINE, of fine quality and very beautiful, has been recently found in Rabun County.

GRANITE and GNESIS. or the best quality for building, abound in the Northern and Middle divisions of the State, and are convenient to transportation.

BUHR rock, from which the best mill-stones are cut, exists, in large quantities, in Burke County ; also in Stewart, Decatur and other parts of the Tertiary formation.

ASBESTUS abounds in most of the Northern counties, and is being mined in the vicinity of Atlanta, Fulton County. Large quantities are regularly shipped to the Northern States, and there worked up in the manufacture of iron safes, fire-proof paints and roofing, lamp wicks, and, to some extent, into cloth. It is being mined with much success, the shipments readily commanding $50.00 per ton.

CALCAREOUS MARLS, or marls composed of shells and other secretions of marine animals, are found in immense beds, in many sections of the State, and in localities where they can be readily utilized for fertilizing purposes. All these deposits exist in the Southern half of the State, or below an elevation of 300 feet above the ocean. It exists in greatest quantities in the Chattahoochee River.

CLAY.—Kaoline, of the finest quality, for the manufacture of porcelain ware, and in the preparation of wall paper, and for other purposes, exists in large and convenient strata in Baldwin and Washington Counties, near the centre of the State, and in Cherokee, Pickens and Union, in the North. Another variety of white clay, suitable for the manufacture of fire-brick, furnace-lining and water-pipes, is also found in Washington and Baldwin Counties, and a large deposit of the same in Richmond. The gray clay used in making pottery, etc., abounds in many counties in the Southeastern portion of the State. Red and yellow clays, suitable for building brick, are found in nearly every county in the State, and in most of them without limit as to quantity.

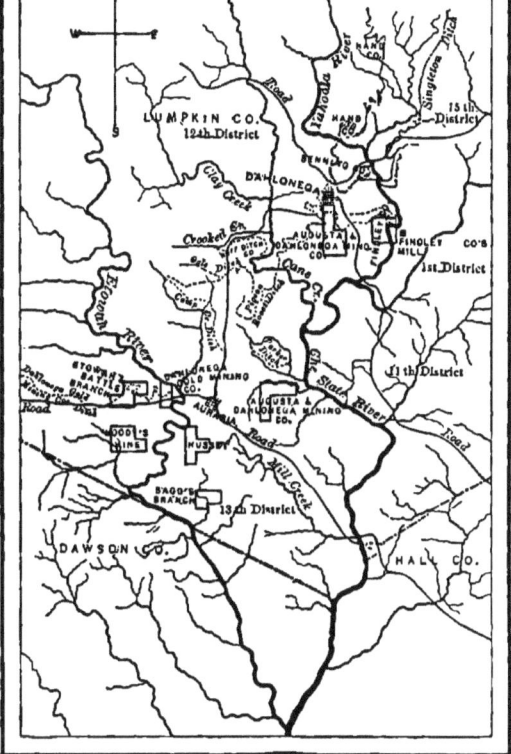

MAP·OF·THE
DAHLONEGA DISTRICT, GA

0   1   2 Miles

N
W · E
S

LUMPKIN CO.
12th District

Nacoola River

HAND

BENNING

DAHLONEGA

15th
District

Singleton Mill

Clay Creek

Crooked Cr.

AUGUSTA &
DAHLONEGA MING
CO.

FINDLEY
MILL

FINDLEY CO'S

1st District

Dane Cr.

Etowah River

Chestatee River

STOWER'S
BATTLE
BRANCH

DAHLONEGA
GOLD MINING
CO.

AUGUSTA &
DAHLONEGA MINING
CO.

11th District

Road

DODD'S
MINE

HUSSEY

BAGG'S
BRANCH

13th District

Mill Creek

Road

DAWSON CO.

HALL CO.

GOLD MINING IN GEORGIA.

When, in 1799, large lumps of gold were found in Georgia, one weighing 28 pounds, exploration was begun, and early in this century the rush to the mines resembled that to California in 1849. In 1830 the State adopted the Indians'* territory and all, and formed Cherokee County, from which many counties have since been made. Hundreds of thousands of dollars have been lost there, and millions made since that date. Hydraulic mining, as conducted in Georgia, is familiar to the reader, probably, but the ease with which the ore is mined, the facility with which it can be sent to market, the cheapness of labor, nearness of timber and water courses, healthfulness, and the ease with which provisions may be procured in Georgia, are not appreciated. Dr. George Little, the State geologist, describes the situation in Georgia as follows:

"It is penetrated by first-class railways, and by short, reliable hack lines. Every part of it is convenient and accessible to cities. Invalids resort there for health. Labor is cheap and plentiful. In California and Arizona it is necessary to transport ore from fifty to three hundred miles; here, transportation is close at hand.

"The greatest advantage, however is, that most of the ore in North Georgia is partly decomposed, and is worked with great facility; where you would have to blast the quartz in California, you can work it with a pick and shovel; consequently ore that is much poorer than California ore can be mined here at a profit, while there it would involve a loss. Besides the above advantages it is very rich, as rich as any ore to be found anywhere."

The following map has been compiled from the latest official surveys and the most authentic sources. It is published by Messrs. Trask & Francis, bankers, 70 Broadway, New York, whose Pocket Mining Atlas has given general satisfaction.

Even those who know the value of the Dahlonega region, in White, Lumpkin and Habersham Counties, are ignorant of the fact that some of the most valuable mines are not in this particular belt. The gold region is much more extensive in Georgia than is popularly supposed. Among the most valuable is probably that known as the "King's Mount Gold Mine," in Hart County, Georgia. This mine is within four miles of the Air Line R. R., is nine miles from Hartsville, the county seat, and thirty-two miles from Athens, Georgia.

The mining property embraces about 502 acres, of which 150 are under cultivation, the rest being heavily timbered.

The estate controls a portion of the water course of the Maria Creek. There are two distinct, continuous veins, according to the report of Mr. F. C. Kropff, of Philadelphia, geologist and mining engineer, that are well defined, productive, and of an advantageous working size. The quartz upon the west vein, at the depth of forty feet, returned, by an assay made by Mr. Kropff, $4.85 per one hundred pounds, while from the average run of the ore taken from this vein out of sump (sixty feet) an analysis made by Dr. Genth, of Philadelphia, gave $3.60 per one hundred pounds, or $71.91 per ton.

Several bushels from the ore-heap out of the fifty-six foot level were reduced by the machinery at the mine, and yielded about two pennyweights to the bushel ($1.92). Mr. Kropff estimated that the total

* There are no Indians in Georgia now.

expenditure for one bushel of one hundred pounds from the miners' hands to tailrace of mill house, amounts to nine and one-half cents, or $1.90 per ton.

After expending about $45,000 for machinery, etc., this mine was abandoned in 1860 by the owners, who reside at the North. The machinery was destroyed during the war; some of the owners have died, and the estate is now in the market.

ACCORDING TO

THE OFFICIAL REPORT OF THE STATE GEOLOGIST.

The gold belt of Georgia is about 100 miles in breadth, with barren intervals here and there. It lies Northeast and Southwest across the entire Northern and part of the Eastern section of the State. It extends through a large number of counties. But few mines have been developed in Columbia and Lincoln Counties, but they are claimed to be among the richest in the State. A vein near Goshen, in the latter county, is said to be yielding at the present time $1,000 per month, at a cost of but $115. While many very rich and profitable mines have been opened in the lower portion of the belt, the greater proportion of the mining has been done in the Northern or mountainous section, especially in the counties of Lumpkin, White Union, Dawson and Cherokee.

" The gold occurs under three distinct conditions: First, as sand (dust), or pebble (nuggets), forming integral portions of the deposits of sand and gravel along the streams, which sometimes extend as high as 100 feet or more above the stream levels. Second, as grains, string or masses, forming integral portions of extensive beds of schists, which are sometimes accompanied by layers of quartz of greater or less thickness, and are sometimes destitute of the least particle of quartz. Third, as a part of the whole of the mineral contents of quartz veins.

" The quartz veins vary in thickness from a few inches to ten feet or more, and have seldom been worked below the water level, from want of capital to purchase the necessary machinery. The ore, when obtained from the veins. is pounded in mills run by water-power, and generally varies from $5.00 to $50.00 per ton, the cost of handling being about 50 cents per ton. There are, however, many instances where the yield has been as high as $60, and even as high as $100 per ton. The business is making steady progress in all the mining districts, and we have returns to date of 34 mills with 537 stamps now in operation, though there are doubtless others not yet brought to our knowledge. The stamps are of hardened iron, and, in weight, range from 350 to 750 pounds. They reduce, each, from one to two tons per day of twelve hours, the quantity depending upon the weight of the stamp and the hardness of the ore. These mills are located chiefly in Lumpkin and White Counties."

Since the above description was penned by Dr. Little, many and valuable improvements have been made, and the most skeptical observer can see abundant evidence of the permanency and value of the mines in this great gold belt.

Messrs. Barlow & Co. have put an additional twenty stamps in the mill attached to the Pigeon Roost Mine, making forty in all in operations on this justly celebrated and well-known mine. The yield has been very satisfactory to the present owners.

THE GOLD REGION – HYDRAULIC MINING AT THE DAHLONEGA MINE.

SMITH BROS.

(From the Report of the State Geologist).

The Hand Canal is about 26 miles in length.   The water is taken from Yahoola Creek, at the foot of the Blue Ridge Mountains.   It is six feet in width at the water line, and four feet in depth ; has a fall of five inches to the 100 feet, and velocity of 30 cubic feet to the second at low water.   At Dahlonega it has an elevation of 220 feet above the Yahoola at this point, and at Findley's Mine a few miles below, it is 300 feet above the level of the Yahoola, which at this point has so enlarged as to be a river.   The reader will form some idea of the power when he contemplates this large body of water foaming along the mountain sides, and ready to be tapped and sent in a resistless torrent, into the vast depths below.   Owing to the rugged nature of the country over which the canal passes, it frequently becomes necessary to conduct this volume of water across immense chasms in order to keep it in its course.   This is done by means of large pipes, which are laid down one mountain side, across the valley, and up the opposite elevation, until it reaches the desired height, and is discharged into a new section of the canal.   There is a pipe near Dahlonega 2,coo feet in length and three feet in diameter.   It is made of boiler-iron strong enough to bear the immense pressure.   There are also on the line of the canal 7,500 feet of wooden tubing, of a like diameter, and secured by strong iron bands.   Between Dahlonega· and the Pigeon Roost mines—the present terminus of the canal—there is another iron tube 2,400 feet in length, and 22 inches in diameter.

The modes of utilizing this water in the operations for gold are various.   It is the motive power of the mills where the stamping and washing are done.   It serves to carry the ores and gold-bearing earth from the mines to the mills, thus saving the greater part of the cost of transportation.   It is also largely used in an operation called " sluicing," where it is turned loose upon the hill-sides and of its own gravity bears away several feet of the surface earth.   For the same purpose a hose and nine inch pipe is sometimes used, and its power in uprooting trees, bearing down mountains and filling up valleys, is truly wonderful.   Often the full force of the canal is turned into a vein containing a day's work of the ore and its rich surroundings, and the whole mass sent roaring down the mountain-side into the mill some thousands of feet below.   Immense boulders of quartz are sent whirling like so many chips or leaves.   This operation not only tears away the earth to the depth of several feet, but at the same time exposes every vein of ore and prepares it for the pick.   Every mill is prepared with a receptacle for these washings, from which the water having been drained off, the ores, gravel and sand are shoved into the troughs and pounded into powder by the immense iron stamps.   The pounded contents are then carried by a stream of water over a copper surface, upon which there is a coating of quicksilver, with which the fine particles of gold form a mechanical union, and from which they are subsequently liberated by the application of heat, the amalgam having been first scraped from the copper sheets and deposited in a crucible. This is the usual process; there are others, but as they differ only in details, it is unnecessary to mention them.

The *Hand Canal* is not only used by the mining company who constructed it, but by all the miners on the line, at a moderate rental paid to the proprietors.   It is said that in this and other improvements, the Hand Company has invested upwards of a quarter of a million dollars.

The hills on both sides of the Yahoola River are gold bearing, and the ore is very rich.   The river furnishes ample water power to work the mines.   Some fifty or sixty additional stamps are to be placed on the Singleton mining property, all of which will have work day and night for many years on ore that cannot fail to pay a big profit on the investment.

A correspondent of the *Mining Record* thus described a few mining properties :

"The Findley mine, since the new machinery has been put in place, is running smoothly along on full time.   A hydraulic pipe has

THE GOLD REGION—DAHLONEGA MILL ON THE ETOWAH.

been attached to the reservoir on the hill and helps materially to break down the masses of ore and slate in the hill, and most signally lessens the labor of getting ore into the mill.

"The Griscomb mill is now preparing to add a pump to throw water on the hill into a reservoir, and from this a hydraulic pipe will be used to break down the ore and slate, and flood it into an ore yard near the mill."

It is manifest to every miner that all the streams cutting this gold belt are rich in many places, and if properly worked, will pay a handsome profit to those engaged in the enterprise.

NOTES ON GOLD MINING IN GEORGIA.

(From the Atlanta Constitution.

The Gold Field.

"ATLANTA, Ga., April 15th.

" Additional advices from the gold field in Nacoochee Valley show richer results than before reported. A solid nugget was found on Monday, weighing 440 pennyweights, without flaw or gravel. Two hands picked up 900 pennyweights in one day. With their ordinary washing-pans and a little stream of water, the total cost of all being $50.75, they took out in nuggets, in four and a half days, 730 penny-weights in gold, worth over $700. In one day they took out $389 worth. In about fifteen days work they have taken out $1,259, and the yield grows richer the further they go. They have worked up so far, a space of only sixteen feet square. The vein is about 100 feet wide, and has been tested for two miles."

The following gold items we take from the Gainesville *Eagle :*

" Mr. Will Logan, of White County, was in the city Wednesday, and had with him a nugget of gold taken from the Richardson mine on Saturday, weighing sixty-five pennyweights. We learn that there is no falling off in the yield of this mine."

In summing up the mineral sources of Georgia, we assert that they are not surpassed by any State in the Union. We have gold rivalling that of California, iron far surpassing that of Pennsylvania, and coal underneath 175 miles of our territory, equal to all our wants. We have copper equal to that of Tennessee, and more easily mined than that used in Michigan. We have manganese five times richer than that used in the Terre Noire Works, France, worth $400 per ton ; and finally, our slate is superior to that in the Lehigh Valley. The min-eral wealth imbedded in the mountains and soils of Georgia is of inestimable value.

VALUABLE MINERALS IN GEORGIA.

Gold is found in 56 counties ; Copper, in 13 counties ; Asbestos, in 12 counties ; Manganese, in 4 counties ; Slate, in 3 counties ; Iron, in 43 counties ; Mica, in 6 counties ; Diamonds, precious stones, gems, etc., in 26 counties. Diamonds are found in Hall and White Counties ; Opal in Bulloch and Washington Counties ; Galena, in 7 counties ; Silver, in 3 counties ; Graphite, in 9 counties ; Kaolin, in 5 counties ; Fire Clay, in 3 counties ; Limestone, in 31 counties ; Buhr-stone, in 27 counties ; Marl, in 29 counties ; Green Sand, in 4 coun-ties ; Marble, in 9 counties ; Gilmer has it white and variegated ; Walker, black marble. Coal, in 3 counties ; Baryta, in 2 counties ;

Serpentine, in 8 counties; Soapstone, in 23 counties; Granite, in 48 counties, in sufficient quantities to be quarried and used for building purposes. Standsone, in 9 counties; Lithographic Stone is found in Walker County. Polishing Sandstone, in 3 counties. Muck, for agricultural purposes, is found in Charlton, Clinch and Wade Counties.

There are a number of other minerals found in different counties. There are a number of others not yet examined by the State Geologist. When there has been a thorough examination of the State, and a map for each county made and marked off, a person can tell at a glance of the eye what minerals each county possesses, and where to find them.

ARE IMMIGRANTS WANTED IN GEORGIA?

To this question the reader is referred to the pamphlet entitled " Georgia," is issued by the State Department of Agriculture. This pamphlet contains letters from every part of Georgia, written by settlers of Northern and foreign birth, affirming that there is no social ostracism or political inequality in Georgia, and that immigrants are cordially welcomed. The following letters will be read with interest by those who contemplate settling in Georgia:

" Hyde Park, Lackawanna County, Pa.

" FRANCIS FONTAINE, Commissioner, etc., 60 East 10th St., N. Y.

" Dear Sir:

" I have read with interest your letter on sheep husbandry in Georgia, as given in the *New York Atlas*, and would like to learn something more definite about land in Georgia, where located, terms on which it can be bought, title, character of the land and locality, and of the people as well, together with any items of interest to one who would wish to settle there.

" I was with Sherman's army in Georgia as far as Atlanta, and have recommended to my friends ever since the war to look at the land in the South before going West, as I believe the South has many advantages over the West to offer to settlers, provided the people are disposed to live in harmony, but the belief is pretty general in the North that the Southern people still feel bitter toward the Northern men, and would make it very unpleasant to live among them. I think that I could live among them without trouble, but what I want is to satisfy others, so that we can organize a colony of emigrants to buy a tract of land and make homes for a number of families. If you have papers that will give the desired information, or if you think proper to write, I will do what I can to distribute the information among working men.   Respectfully yours,

" CHARLES CORLESS."

VIEW IN CENTRAL CITY PARK, MACON, (MIDDLE GEORGIA.)

A PROSPEROUS NORTHERN SETTLEMENT.

"MADISON, GA., February 16, 1880.

"*Editors Chronicle and Constitutionalist:*

"You will find enclosed a letter, descriptive of the agricultural attractions of our section, written by F. C. Foster, of our city. It is not generally known, although it is true, that Morgan County, more than any other county of its size in Georgia, except such as may contain a large city, has succeeded in inducing immigration. I suppose not less than half a million dollars of money have been invested in property in our county by Northern settlers. Our example might interest and stimulate other counties where your paper circulates, and for this reason I submit the enclosed letter for publication.

"Respectfully,

H. W. BALDWIN."

"MADISON, GA., January 19, 1880.

"DEAR SIR:—It has long been my desire that the people of the North should know something of this section, hence I will avail myself with much pleasure of your kind invitation to use your columns, only regretting that an abler pen should not have been called upon to describe this county and this people.

"I shall attempt nothing but a plain statement of facts, which are verified by the endorsements attached of men of intelligence and character, who have moved from the North, and settled here since the war.

"This county (Morgan) is one of the central counties of the State—in the very heart of Middle Georgia. Topography, gently undulating, abundantly watered with branches, creeks and rivers. A splendid red clay subsoil, originally covered with grand forests of oak, pine and hickory, large tracts of which are still in their pristine fertility, not unequal to any other portion of this great continent. Seven hundred and sixty feet above the sea, with a climate unsurpassed on earth, a generous soil adapted to the growth of corn, cotton, wheat, oats and all other small grain, the finest grasses growing spontaneously during the greater part of the year. A superabundance of clear, pure free-stone water—the creeks and rivers dashing over shoals furnish a boundless supply of water-power. There is and can be no good reason why all the various industries and enterprises which go to make a country prosperous and the citizens happy and contented, should not flourish here. There is one cotton factory in the county run by water power, which has been in operation continuously for the past ten or fifteen years. It is owned by a company composed, as I am informed, of a very few members; and while it is fifteen miles from the nearest railway depot, the fact that none of its stock is or has been for years on the market, is an evidence of its prosperity. Others might be constructed with even more water-power within two miles of the trunk line of the Georgia Railroad.

"The population of the county is between 10,000 and 12,000, among which quite a number of Northern settlers who have purchased land since the war and engaged in farming, fruit culture, cattle and sheep raising. Some of the settlers inform me that the Bermuda grass, which grows luxuriantly without any culture or attention whatever, is a finer grass for hay or pasturage than any they have ever seen.

Madison, the county seat, is a healthful little city of 3,000 souls, situated on the Georgia Railroad, 78 miles from Atlanta, and 104 miles from Augusta   In this little city there are four churches for whites – Baptist, Methodist, Episcopal and Presbyterian—and two colored churches – Baptist and Methodist.   A female seminary and . well organized male academ , besides schools for smaller children under competent and experienced teachers, flourish in our midst.

"In this connection it may not be amiss to state that Georgia is far in advance of any other Southern State in her efforts in behalf of popular education.

"The relations existing between the Northern settlers who came here since the war, and the natives, are of the most cordial and social character.   Recognizing the fact that they are citizens of one common country, they live together in perfect harmony, rendering mutual aid and assistance to each other.   Some of these settlers the writer numbers among his warm personal friends.   As an evidence of how they are received in our midst, I will state that one of them, Mr. E. Heyser, without in the least seeking the office, was elected by the people to the honorable and responsible office of Clerk of the Superior Court, which office he now fills with credit to himself and to the perfect satis-faction of his fellow citizens.   Another, Mr. John H. Morgan, has been twice elected by the grand jury of the county into the board of county commissioners, where he, as its chairman, has faithfully and honorably served, in connection with his colleagues, in looking after the county business and managing its finances   Both of these gentle-men, I am informed, are, in national politics,. Republicans, but their political principles had nothing whatever to do with their election to the offices now filled by them.

" The people are intelligent and law abiding, and the honorable men whose names are attached to the endorsements hereto will all bear testimony to the fact that any man, of whatever country, of whatever religious or political caste, can come here and proclaim his sentiments, principles and political opinions as free as air, without let or hindrance, and that no man has or ever will be interfered with in the enjoyment of any of his rights or privileges as a citizen.   But on the contrary, a cordial invitation is extended, and a cordial welcome will be given to any one wishing to settle here, and cast his lot with ours.

"If any of your readers should desire information concerning this section, I or any gentlemen whose names are appended will cheer-fully answer any communication addressed to either one of us.

"Thanking you for the use of your columns, and your uniform kindness to me, I am yours truly,

"F. C. FOSTER."

" I carefully endorse the statement stated by Mr. F. C. Foster. I have traveled over most of the United States, and will candidly say that I prefer this section to any other. I came here from Western New York, five years ago, and myself and family are highly pleased with the hospitality and kindness extended to the Northern emi-grants, while the climate is all that can be desired, and combined with the great variety of fruits that can be raised, in connection with good water, and being healthy, with quick running streams, makes it all that heart can wish.   "H. C. ADGATE."

"Having carefully examined the above article setting forth the agricultural, climatic, educational and other advantages of Morgan County, Georgia, I fully endorse every statement made in said article by the Hon. F. C. Foster, as true and correct. After a residence in this county for seven years, I have no desire to return to the place of my nativity in Eastern Pennsylvania.

"ALLEN HEYSER."

"I am a native of Scotland; came to Philadelphia in 1868; removed to this county in 1871. I endorse every statement made in letter of Mr. F. C. Foster, and I intend to live here the balance of my life, being better pleased with this county than any I have ever seen.

"JAMES W. MCMILLAN."

"I came to Morgan County, Ga., in 1876, from Evansburg, Pa. I endorse the statement made in Mr. Foster's letter to the fullest extent.

"E. HEYSER, Clerk Superior Court."

"MADISON, GA., January 19, 1880.

"I came to Madison from Paterson, N. J., in 1870; in 1877 married into a Southern family. Do most cheerfully endorse the article of F. C. Foster, and assert that I am delighted with the people, the climate and the State.

"A. K. AKERMAN."

"I came from Poughkeepsie, N. Y., to Morgan County, Ga., in 1874. Purchased land here, and am now engaged in grain raising and fruit culture. Have lived in San Francisco, and consider this the best country and blest with best climate I have ever known. I endorse Mr. Foster's letter in the very fullest extent.

"E. D. MULFORD."

"We came to Morgan County, Ga., from Wyoming County, N. Y., in 1871. Purchased land here, and are now engaged in farming. This is a fine grain and grass country; magnificent climate. Our relations with the natives are of the most cordial character. We would not return to the farm left by us in New York if it were given to us. We are Republicans in politics.

"J. M. GRIGGS,
"P. M. GRIGGS."

"I came to Morgan County, Ga., from Columbia County, N. Y., January, 1866. Purchased land and am now growing grain, cotton, raising Angora goats, fattening cattle. I have fifteen apiaries, from which I extracted last year between 4,000 and 5,000 pounds of honey; I have twenty orchards, raising as fine peaches as are growing on earth, and in great abundance from June to the middle of October; there is no better grape country. Too much could be hardly said of Bermuda grass for hay and pasturage, and it grows luxuriantly without any attention. I have always been treated as kindly and hospitally by the natives as if I was born and raised here. I am Republican in politics to the backbone.

"REUBEN MILLER."

"Below I send you a list of settlers who came to this country since the war.   Most of them are free holders and all are citizens.
.   "Yours, truly,
                                                "F. C. FOSTER."

"John H. Morgan, A. B. Daniels, J M. Daniels, of Wisconsin; Reuben Miller, Wm. Washburn. C. P. Kirley, Wm. Smith, Mrs. Wm. H. Crawford, Samuel N. Copeland, Albert J. Howell, John M. Howell, J. L. Howell, Wm. H. Hills, Jasper M. Griggs, Philip M. Griggs, J. A. Valance, John M. Boughton, J. Webster, James Boughton, Wm. Cole, Martin Clark, H. C. Adgate, Frank Hensler, Mrs. M. Bloss, E. D. Mulford, Charles Reed, of New York State; J. M. Roseneranz, S. O. Wilson, John M. Van Winkle, Q. Waathy Van Winkle, John G. Ackerman, of New Jersey; A. Houseman, J. Howe, J. M. Alister, John Heyser, Allen Heyser, Emanuel Heyser, of Pennsylvania; James H. Ainslie, Alex. Monro, John Huff, Mrs. H. Ames, William Tibbald, of Ohio; Jacob Haag, George Haag, of Germany; James W. McMillan, Scotland; P. B. Woodward, of Connecticut.

STATE OF GEORGIA, ⎫ ss. :
  Morgan County.   ⎭

"I, Emanuel Heyser, Clerk of the Superior Court, in and for said county, do hereby certify that the above and foregoing comprise a list of persons who have moved into said county since the close of the late war, and that the above named are now residing in said county, most of whom are heads of families and have purchased lands.

"Given under my hand and seal of ⎫
    Court, this January 20, 1880, ⎭
                            "E. HEYSER,
                                "Clerk, S. C."

GENERAL ADVICE.—HOW TO GO TO GEORGIA.

The immigrant who is absolutely destitute, and who intends to become a laborer on a farm for hire, should not come to Georgia unless arrangements have already been made for his employment.   If he has enough money to pay his transportation, which will cost about twelve ($12) dollars from New York City to any part of the State, and enough money to maintain himself until employed is secured, he need not hesitate.   The people desire that the immigrant shall prosper, and shall become the owner of the soil he cultivates.   It must be borne in mind that nearly all the colored population are agricultural laborers, and hence the demand for farm laborers is not great at present.   Mechanics and skilled laborers will probably get remunerative employment immediately, but all should have some money to begin life in their new homes with.   This will insure independence while seeking to locate one's self.   The immigrant who desires to settle in Southern Georgia should purchase a through ticket *via* Savannah, Ga., by the Georgia Central Steamship Line—cost of steerage passage, $10—or *via* Charleston, S. C., by the New York & Charleston Steamship Line.   The officials at Castle Garden will direct him how to procure tickets and check baggage to its destination.   No other agents, except the accredited agent of the Commissioner of Land and Immigration of Georgia should be listened to, or the agents of the railroad and steamship lines.   The immigrant referred to above will easily find the steamship and railroad offices of the lines leading to Georgia by enquiry at Castle Garden.   In no case seould tickets for Georgia be bought in Castle Garden, for no authority from any railroad or steamship line leading to Georgia has been granted to any one to sell tickets in Castle Garden.

If the immigrant desires to settle in Northern Georgia, he will purchase his ticket from the East Tennessee, Virginia, & Georgia Railroad, No.   Broadway, New York.   The fare by this route, including the steamship fare to Norfolk, Virginia, and thence by rail to Dalton, Georgia, is ten dollars ($10).   If he desires to go to Northeastern Georgia, he can go to Atlanta, and thence to any point by the same route; or, by paying sixteen dollars ($16), he may go from New York to Atlanta *via* the Piedmont Air Line Railroad, office, No. 9, Astor House, New York.   Most of the railroads throughout the State of Georgia will pass immigrants at one (1) cent a mile.

Cheap freight rates on household goods and utensils have been secured from New York to Savannah, and thence by the railroads. Most of these things can be bought in Georgia at such prices that it is not deemed advisable to pay freight on them to Georgia.

The cost of living in Georgia is cheaper than in the West.   The droughts of Colorado, New Mexico and Texas are unknown in Georgia, and the State is not troubled with insects which so ravage Kansas and other Western States.   In comparison with Montana and Minnesota, Georgia offers five months more in which to labor during the year.   Her early fruits and vegetables are among the first in the New York markets, and no State in the Union can better claim a climate which will admit of field work every week in the year.   Ice is manufactured and sold in Georgia at a price less than that charged for it in New York, eight pounds being sold for five cents.   Breweries have been established at Atlanta and other places, and the beer made is of best quality.

Immigrants who speak a foreign language only should come, if possible, in colonies, in order that they may have at all times congenial society.   Write to the Commissioner of Immigration, Atlanta, Georgia, for terms concerning land, water-power, manufacturing sites, prices in Georgia, etc., etc., and know before you come where you are going, cost of transportation, and all necessary facts. Enclose stamp in letter.

To the Northern people, especially the soldiers in the late war between the States, we extend a cordial invitation to become citizens of Georgia. Georgians fought you bravely in war, but that is over, and we welcome you as sincerely in peace. We reverence the memory of our own heroic dead, who fought and died in defence of what they esteemed their constitutional rights, but we bear no malice toward those who as honestly differed with us.

> "The soldier's spirit greets the soldier's call,
> There is no hate between the brave and brave;
> And he whose hand in battle labored first,
> When darkness falls will labor first to save."

With patriotic devotion unexcelled in history, the Southern people in defence of the rights of the States, freely contributed their fortunes and the lives of the flower of the population—the wealthiest bearing the brunt of the conflict as private soldiers, side by side with the poorest, from Manassas to Appomattox, when General Grant returned the sword of General Lee with the chivalric remark: "You are overcome, not conquered." But we accept, without murmuring, the results of that conflict. The amendments to the Constitution have been honestly and finally accepted; secession is no longer claimed as a constitutional right, and there is not one man in ten thousand in the South who would restore slavery if he could. The incubus of slavery has been removed, finally and forever, and the civil and political rights of the lately enfranchised Negro is, in Georgia, as perfectly protected and maintained as in Massachusetts or any State in the Union. The gallant deeds of the American soldier, South and North, is a common heritage of all Americans.

APPENDIX.

The appendix added here contains " A Partial List of Water-Powers in Georgia," which was prepared by Dr. Little, State Geologist, for the Hand-Book of Georgia, published by Dr. Janes, the former Commissioner of Agriculture. Also List of the Woody Plants of Georgia, which was also prepared by Dr. Little. For the Statistical Tables added, I am indebted to Chief Nimmo of the United States Bureau of Statistics. It is regretted that the Census returns for 1880, are not completed before this little work goes to press. Land Tables, giving the price, location and characteristics of land offered for sale in every part of the State, will be prepared as soon as circumstances will permit.

A PARTIAL LIST OF THE WATER-POWERS IN GEORGIA, WITH DESCRIPTIONS, ARRANGED BY COUNTIES.

Name of Stream.	Point of Section.	Cubic feet per second.	Theoretical horse-power of one-foot head.	Available horse-power of one-foot head.	Approximate head or an assumed head of 10 feet.	Theoretical power of stream with this head running 24 hours.	Available power of stream with this head working 24 hours of each day.	Condition of stream.	By whom surveyed.	Remarks.
Banks County.										
Broad River..........	Habersham Line.......	27.20	3.10	2.48	10.00	31.00	24.80	Low water or more.	Barrow and Locke.	Water very low.
Grove River..	Homer and Mt. Airy Road..	65.60	7.41	5.92	10.00	74.10	59.20	" "	Locke.	" " "
Hudson River....... ...	" " " " ...	77.40	8.59	6.86	10.00	85.81	68.61	" "	"	" " largest spring in county.
Barrow County.										
Oothcaloga Creek.....	Gordon Line.......	15.00	1.70	1.36	6.00	10.20	8.16	Minimum low water.	"	Water very low.
" "	Adairsville.....	7.00	.79	.63	6.00	5.36	4.27	"	"	" " "
Lewis Spring........	Near Adairsville....	8.00	.80	.64	10.00	9.12	7.3	"	"	" " "
Cedar Spring........	Martello's Mill........	2.50	.28	.22	18.00	5.10	4.0	"	"	Water very low.
Cedar Creek........	Gordon Line........	8.00	.80	.64	12.00	11.00	8.0	"	"	" " "
Fork of Pine Log......	McCanless and Parrott Mill.	18.00	2.04	1.63	20.00	41.00	32.8	"	"	" " "
" " "	Johnson's Mill.......	14.00	1.60	1.28	15.00	24.00	19.2	"	"	Estimated.
Silacoa Creek........	Gordon Line........	20.00	2.27	1.81	20.00	45.60	36.5	"	"	Very low.
Stamp Creek.........	Pool's Furnace.......	12.00	1.34	1.07	20.00	27.3	22.0	"	"	" "
" "	At mouth.......	24.00	2.68	2.14	20.00	54.4	43.7	"	"	" "
Boston's Creek.	" "	4.00	.45	.36	20.00	9.2	7.3	"	"	" " very rapid fall

Rogers Creek	At mouth	7.00	.79	.63	20.00	16.00	13.00	Low water.	Locke.	
Etowah River	At mouth of Allatoona	1307.7	147.68	118.14	15.00	2250.00	1835.00		"	
Pettis Creek	Mouth	20.00	2.67	2.13	5.00	12.00	9.60	Minimum low water.	"	Very low.
Nancy Creek	"	6.00	.68	.50	5.00	3.00	.....	Minimum low water.	"	
Two-Run Creek	Kingston	26.00	2.94	2.3	16.00	46.00	38.40	Low water.	"	
Conaseena Creek	"	5.00	.55	.44	20.00	11.00	9.10	"	"	
Baresley's Creek	Near mouth	5.00	.55	.44	18.00	10.00	8.2	"	"	
Allatoona Creek	2½ miles from mouth	25.5	28.50	22.80	17.00	46.4	38.8	"	"	
Pumpkinvine Creek	2 " " "	70.00	7.95	6.41	10.00	80.00	64.0	"	"	
Raccoon Creek	1 " " "	39.00	4.54	3.03	10.00	45.00	36.5	"	"	
Euharlee	2 " " "	120.90	13.51	10.81	12.00	165.6	132.8	"	"	
BIBB COUNTY.										
Ocmulgee River	Holt's Shoals	2917.00	331.37	205.09	3.70	1224.70	979.76			
Walnut Creek	Macon	5.00	.57	.45	10.00	5.70	4.50	" "	"	Estimated.
Swift Creek	7 miles, Macon	5.00	.57	.45	10.00	5.70	4.50	" "	"	"
Stone Creek	8 " "	8.00	.91	.72	12.00	10.92	8.73	" "	"	
Tobesofkee Creek	Freeman's Mill	70.00	7.98	6.38	20.00	159.60	127.68	Above " " "		
BURKE COUNTY.										
McBean's Creek	Waynesborough R.R.	50.00	5.70	.56	10.00	570.00	456.00		Barrow.	Low flat banks.
Boggy Gut Creek	Shell Bluff	10.00	1.14	.91	10.00	11.40	9.10		"	"
Sapp's Spring Creek	Sapp's Mill	20.00	2.28	1.82	11.00	25.08	20.06		"	
CARROLL COUNTY.										
Buffalo Creek	1¼ miles south of Carrollton	6.00	.68	.51	10.00	6.80	5.40		Locke.	
Briar Creek	3 miles, Carrollton	5.00	.56	.44	10.00	5.00	4.40		"	
Panther Creek	4½ " "	4.00	.45	.36	10.00	4.50	3.00		"	
Buffalo Creek	1 mile above mouth	18.00	2.01	1.64	10.00	20.40	16.40		"	Estimated.

A PARTIAL LIST OF THE WATER-POWERS IN GEORGIA, ETC.—(continued.)

Name of Stream.	Point of Section.	Cubic feet per second.	Theoretical horse-power of one-foot head.	Available horse power of one-foot head.	Approximate head or an assumed head of 10 feet.	Theoretical power of stream running with this head 24 hours.	Available power of stream with this head working 24 hours of each day.	Condition of Stream.	By whom surveyed.	Remarks.
CARROLL CO.—*Continued.*										
Snake Creek	Factory	42.00	4.70	3.76	30.00	141.00	112.80	Low spring	Locke.	100 or more feet of head can be had.
Dog River	Above Watkins' Mill	25.76	2.92	1.60	10.00	29.20	16.00	"	"	Measurement unsatisfactory.
Cockrum's Creek	Old Cherokee and Carroll Line	4.5	.60	.48	10.00	6.00	4.80	"	"	Estimated.
Tallapoosa	Above mouth of Buck Creek	101.43	17.42	9.13	10.00	114.20	91.30	"	"	
Buck Creek		16.60	1.81	1.45	10.00	18.10	14.50	"	"	
Indian Creek	South of Tallapoos and near Bonner's	7.00	.79	.64	10.00	7.91	6.40	"	"	Estimated.
Whooping Creek	Dorris Mill	24.50	2.72	2.17	10.00	27.20	21.70	Flush or less.	"	
CHATTAHOOCHEE CO.										
Oewitchee Creek	Bagley's Mill	6.00	.70	.56	18.00	12.60	10.08	Low spring.	"	
"	Romney's Mill	21.00	2.98	1.82	18.00	33.04	24.43	"	"	
Woolfolk's Branch	Woolfolk's	1.00	.11	.08	65.00	7.15	5.72	"	"	
Upatoi		....	....	....	12.00	....	....	"	"	Very sandy and full.
CHATTOOGA COUNTY.										
Little Turtle Creek	Near mouth	5.5	0.62	0.49	10.00	6.20	4.90		Barrow,	

Stream	Locality									Remarks
Raccoon Creek	Lot 39	4.5	0.51	0.40	10.00	5.10	4.00		Barrow.	
Rough Creek	Mouth	8.8	1.00	0.90	10.00	10.00	8.00		"	
Armuchee Creek	Subligna	41.5	4.73	3.78	10.00	4.73	3.78		"	
CHEROKEE COUNTY.										
Mill Creek	Mouth at Canton	46.00	5.22	4.17	10.00	52.20	41.70	Low spring or more.	"	Cubic feet estimated.
CLAY COUNTY.										
Chemochechobee	Weaver's Mill	60.00	6.84	5.47	30.00	205.20	164.16			
Pataula	Rapids	240.00	27.36	21.88	22.00	601.92	481.53			
CLINCH COUNTY.										
Suwanee River	Mixon's Ferry	72.00	7.95	6.38	10.00	79.5	63.80	Minimum low water.	Locke.	
COBB COUNTY.										
Big or Vickery's Creek	Empire Mill	147.	16.76	13.4	16.0	268.1	214.52	Low water.	Col. Robinson, R. M. Co.	
" "	Roswell Manufacturing Co	147.	16.76	13.40	30.10	502.50	402.24	"	"	
" "	Lebanon Mills	147.	16.76	13.40	14.00	234.6	187.7	"	"	
Head of Nickajack	Jones' Mills	3.00	0.34	0.27	15.00	5.10	4.08	Low spring.	Locke.	
Nickajack	Ruff's Mills	....	....	....	29.00	....	....			
"	Concord Factory	....	....	....	21.00	....	....			Too full for measurement, has probably 20 cubic feet at low water.
"	Concord Factory and Ruff's Mill combined	....	....	....	50.00	...	...			
Chattatoochee	Austell's Shoals	2000.00	226.20	180.96	10.00	2262.00	1809.60		"	
Tributary Sweet Water	Babb's Mill	2.00	0.23	0.18	18.0	4.14	3.32		"	Estimated.
Rotten Wood	Aker's Mill	35.00	3.97	3.17	32.0	127.24	100.78	Low water.	"	Almost any head to 50 obtainable.
" "	Boring's Mill	38.00	4.30	3.44	10.00	43.00	34.40	Low spring.	"	

A PARTIAL LIST OF THE WATER-POWERS IN GEORGIA, ETC.—(continued.)

Name of Stream.	Point of Section.	Cubic feet per second	Theoretical horse-power of one-foot head.	Available horse-power of one-foot head.	Approximate head or an assumed head of 10 feet.	Theoretical power of stream with this head running 24 hours.	Available power of stream with this head working 24 hours of each day.	Condition of Stream.	By whom surveyed.	Remarks.
COBB CO.—*Continued.*										
Soap Creek...........	At Paper Mill	62.00	7.40	5.92	67.00	495.8	396.64	Low spring.	Locke.	Head includes Robertson's Mill.
Little Willico......	Old Starch Factory....	5.00	.57	.45	20.00	11.40	9.00	"	"	There are two L. Willicos
" "	At mouth, Willico Factory..	8.00	.908	.72	30.00	27.00	21.60	" or more	"	" " " "
Willico...........	Above Factory....	21.60	2.45	1.96	31.00	75.95	60.72	Low spring.	"	
Powder Spring Creek.	Powder Spring....	34.00	3.96	3.17	10.00	39.60	31.70	"	"	
Sweet Water........	Hays' Bridge....	80.50	9.00	7.20	10.00	90.00	72.00	Low water.	"	
COLUMBIA COUNTY.										
Kiokee Creek.	Near Appling....	30.00	3.42	2.73	10.00	34.20	27.30		Barrow.	
DAWSON COUNTY.										
Etowah River....	Palmer's Mill....	60.25	6.87	5.29	10.00	48.70	52.90		"	
Shoal Creek........	Howzer's Mill....	33.00	3.76	2.86	16.00	60.16	48.12		"	
Amicolola River......	Dawsonville and Jasper Road.	163.60	11.80	9.44	51.00	590.00	472.00		"	
" "	8 miles Dawsonville....	85.00	9.69	7.75	10.00	96.96	77.50		"	
Head of Jones' Creek..	Foster's Mill....	2.00	.23	.18	14.00	3.19	2.55		"	

DECATUR COUNTY.										
Limesink	Limesink	2.00	0.23	0.18	105.00	24.15	19.32	Low spring.	Locke.	Creek disappears. Probably has more water.
Barnet's Creek	Lot 367	23.00	2.62	2.09	10.00	26.20	20.90	"	"	Flow affected by mills above.
Attapulgus Creek	Thomasville Road	18.00	2.05	1.64	10.00	20.50	16.40	"	"	Estimated.
Martin's Mill Creek	"	5.00	0.57	0.45	7.00	3.99	3.19	"	"	
Sanburn's Mill Creek	Attapulgus Road	8.00	0.91	0.72	10.00	9.10	7.20	"	"	
DE KALB COUNTY.										
Peachtree Creek	Houston's Mill	23.75	2.71	2.16	22.0	.....	.....	Low water.	"	
EARLY COUNTY.										
Harrod's Creek	Early Factory	20.00	2.28	1.82	35.00	79.80	63.84	Low spring.	"	
Colomochee Creek	Early Road	70.00	7.98	6.38	12.00	95.76	76.60			Estimated.
ELBERT COUNTY.										
Beaver Dam Creek	E. A. L. R. R.	30.00	3.42	2.73	10.00	34.30	27.30		Barrow.	
FLOYD COUNTY.										
Armuchee Creek	Jones' Mill	135	15.40	12.32	10.00	151.3	121.04	Low water.	Locke.	Stream a little above L. w.
Lit. Fork Armuchee Ck.	Texas Valley Road.	41	4.67	3.73	15.00	71.1	55.95	"	"	"
Big Fork Armuchee Ck.	" "	48	5.47	4.37	10.00	54.7	43.7	"	"	"
John's Creek	Near mouth	15	1.71	1.36	8.00	13.6	10.88	"	"	"
Silver Creek	" "	24	2.73	2.18	18.0	49.2	39.24	"	"	"
Cedar Creek	Thoman's Mill	70	8.00	6.40	10.0	79.8	64.0	Minimum low water.	"	"
Little Cedar Creek	Near month.	20	2.28	1.82	14.0	31.9	25.50	Low spring or more.		
" " "	Cave Spring.	60.80	6.92	5.54	10.00	69.20	55.41	" "		
Big Spring	" "	7.98	.90	.786	10.00	0.08	7.86	Low spring.		

A PARTIAL LIST OF THE WATER-POWERS IN GEORGIA, ETC.—(continued.)

Name of Stream.	Point of Section.	Cubic feet per second.	Theoretical horse-power of one-foot head.	Available horse-power of one foot head.	Approximate head or an assumed head of 10 feet.	Theoretical power of stream with this head running 24 hours.	Available power of stream with this head working 24 hours of each day.	Condition of Stream.	By whom surveyed.	Remarks.
FORSYTH COUNTY.										
Beaver Run	Mouth	75.00	8.56	6.84	20.00	171.00	136.80	Flush.	Barrow.	
Sitting-Down Creek	Holbrook's Mill	30.00	3.42	2.73	7.00	23.94	19.15	Low spring.	"	
Etowah River	Franklin Mines	1129.00	128.70	102.96	8.00	1029.00	823.20		"	
Sitting-Down Creek	Pool and Heard's Mill	30.00	3.42	2.73	15.00	51.70	41.36		"	
FRANKLIN COUNTY.										
Broad River	Toccoa and Carnesville Road	50.00	5.70	4.56	10.00	57.00	45.60		"	
Creek	3 miles Carnesville	2.00	0.23	.18	16.00	3.65	2.92		"	
Unawattee	4 " "	50.00	5.70	4.56	10.00	57.00	45.60		"	
FULTON COUNTY.										
Peachtree	Atlanta and Buckhead Road	97.50	11.07	8.85				Flush or lower.	Locke.	
Nancy's Creek	96 and 17	45.00	5.01	4.01				Low spring.	"	
Marsh Creek	73 and 17	5.00	0.57	.45				"	"	Estimated.
Long Island Creek	164 and 17	6.5	.73	.58				"	"	"
GLASCOCK COUNTY.										
Sock's Branch	Mouth	6.00	0.68	0.54	18.00	12.31	9.84			

GORDON COUNTY.									
Oothcaloga	Calhoun Mills	41.36	4.71	3.76	9.0	42.39	33.91		Barrow.
Connesauga	Mouth	293.0	32.10	25.68	10.00	321.00	256.80		"
Craneta Springs	5 miles Calhoun	6.00	0.68	0.54	12.0	8.2	6.56		"
Smoke Creek	Near mouth	5.00	0.57	0.45	10.00	5.70	4.50		"
Coosawattee	Carter's Mill	541.0	61.70	49.36	50.0	3065.0	2468.0		"
Talking Rock	At month	107.90	12.30	9.76	10.00	122.00	97.60		"
Dry Creek	Lot 85	8.00	0.91	0.72	10.00	9.10	7.20		"
Salacoa	117, 7, and 3	119.6	13.63	10.90	10.00	136.30	109.00		"
Resaca Creek	Resaca	12.40	1.41	1.12	10.00	14.10	11.20		"
Lick Creek	Lot 116	6.00	0.68	0.54	10.00	6.80	5.40		"
Snake Creek	113 and 1	14.70	1.67	1.33	10.00	16.70	13.30		"
Rocky Creek	14, 24, and 3	3.50	0.39	0.31	10.00	3.90	3.10		"
John's Creek	53, 24, and 3	12.56	1.43	1.14	10.00	14.3	11.40		"
GWINNETT COUNTY.									
Yellow River	Fain's Mill	60.0	6.84	5.47	20.0	136.8	109.4		Barrow and Locke. Estimated April 24th for low water.
" "	Stedman's Mill	64.00	7.30	5.84	30.00	219.0	175.20		" "
" "	Montgomery's Mill	38.40	4.38	3.50	14.00	61.32	49.00 Low spring		" " Or higher.
Wolf Creek	Near Montgomery's Mill	5.00	0.57	0.45	10.00	5.7	4.5	"	" "
Suwanee Creek	Lawrenceville and Buford Rd	11.85	1.34	1.07	10.00	13.40	10.70	"	" "
Level Creek	Strickland's Mill	12.00	1.36	1.08	20.00	35.44	28.35	"	" "
Ivy Creek	Hamilton's Mill	2.00	0.23	0.18	18.00	4.10	3.28	"	" "
HABERSHAM COUNTY.									
Hazell Creek	Clarksv'le and Gainesv'le Rd	31.85	3.60	2.88	8.00	28.80	23.04 Above l.w.		" "
Soquee River	Clarksville	124.86	13.74	10.99	10.00	137.40	109.90	"	" "
Sheal Creek	Crow's Mill	3.0	0.34	0.27	12.00	4.10	3.28	"	" "
Tallulah River	Above Falls	458.5	51.27	41.01	400.0	20508.00	16406.40	"	" "

A PARTIAL LIST OF THE WATER-POWERS IN GEORGIA, ETC.—(continued.)

Name of Stream.	Point of Section.	Cubic feet per second.	Theoretical horse-power of one-foot head.	Available horse-power of one-foot head.	Approximate head or an assumed head of 10 feet.	Theoretical power of stream with this head running 24 hours.	Available power of this head working 24 hours of each day.	Condition of Stream.	By whom surveyed.	Remarks.
HABERSHAM Co.—Cont.										
Panther Creek	Weaver's Mill	19.37	2.22	1.76	30.00	66.66	53.22	Low water.	Barrow andLocke	Falls rapidly.
Rock Hazel Creek	Jackson's Mill	3.00	0.34	0.27	20.00	6.80	5.40	"	"	
Mud Hazel Creek	Near mouth	8.85	1.00	0.80	10.00	10.00	8 00	"	"	
Little Mud Creek	½ mile Hall Line	38.00	3.76	3.00	10.00	37.60	30.00	Above l. w.	"	
Big Mud Creek	" " "	20.00	2.28	1.82	10.00	22.80	18.20	"	"	
Ward's Creek	Jarrett's Mill	33.75	3.76	2.56	10.00	37.60	28.60	Flush.	"	
Toccoa Creek	Toccoa Falls	5.20	0.60	0.48	190.00	114.00	91.20	Low spring.	"	
Roper's Creek	Willbank's Store	5.00	0.57	0.45	10.00	5.70	4.50	"	"	
Soquee River	Hill's Mill	41.04	4.60	3.68	40.00	184.00	147.20	"	"	
Sutton's Mill Creek	Near Clarksville	16.80	2.00	1.60	10.00	20.00	16.00	"	"	
Deep Creek	Near mouth	38.50	4.89	3.51	10.00	43.90	35.10	"	"	
Creek	Near Batesville	3.00	0.34	0.27	9.00	3.07	2.45	"	"	
Mathews' Mill Creek	Mouth	1.50	0.17	0.13	22.00	3.76	2.86		Barrow.	
Panther Creek	Walker's Mill	4.50	0.51	0.40	20.00	10.26	8.20	"	"	
Nancy Town Creek	At mouth of Cox's Creek	5.29	0.60	0.48	10.00	6.00	4.80	"	"	
Cox's Creek	Near mouth	2.00	0.22	0.17	100.00	22.00	17.60	"	"	
Nancy Town Creek	Above Slack's Branch	2.80	0.32	0.25	15.00	4.78	3.88	"	"	
Dick's Creek	Hulsey's Mill	3.32	0.37	0.29	30.00	11.30	9.04	"	"	
Leatherwood Creek	Hickey's Mill	0.75	0.08	0.06	14.00	1.30	0.96	"	"	

Walton's Creek.......	Jarrett's Bridge Road......	5.10	0.58	0.46	10.00	5.80	4.60	Barrow.
Toccoa Creek........	At mouth............	16.00	1.82	1.45	10.00	18.20	14.50	"
Black Mountain Creek.	Near mouth	1.25	0.14	0.11	10.00	1.40	1.10	"
Panther Creek........	" "	53.63	6.11	4.68	10.00	61.10	48.80	"
HALL COUNTY.								
Chestatee............	Leather's Ford........	290.00	33.00	26.40	12.00	396.00	316.80	"
Yellow Creek........	Near mouth..........	7.28	0.83	0.66	20.00	16.60	13.28	"
Big Wahoo Creek.....	Glade Mine and Leatherwood Ford Road.........	14.57	1.66	1.32	10.00	16.60	13.20	"
Middle Wahoo Creek...	Glade Mine and Leatherwood Ford Road..........	12.47	1.42	1.13	10.00	14.20	11.30	"
Little River.........	Glade Mine and Leatherwood Ford Road.........	12.64	1.44	1.15	10.00	14.40	11.50	"
Flat Creek...........	Above Glade Mine......	17.28	1.97	1.57	50.00	98.50	78.80	"
Chattahoochee River...	Shallow Ford........	929.00	106.00	84.80	10.00	1000.00	848.00	"
North Fork Oconee.....	Sulphur Springs......	22.37	2.54	2.03	10.00	25.40	20.30	"
" " "	Carnesv'le and Gainesv'le R'd	31.50	3.59	2.87	10.00	35.90	28.70	"
Candler's Creek.......	Mouth..............	9.60	1.10	0.88	10.00	10.9	8.80	"
Pigeon-Wing Creek....	Mouth..............	2.00	0.23	0.18	10.00	2.30	1.80	"
Caney Fork..........	County Line.........	12.00	1.37	1.11	10.00	13.70	11.10	"
Walnut Fork.........	Harrington's Ford.....	13.54	1.77	1.41	20.00	35.40	28.32	"
Holly Branch........	Mouth..............	2.50	0.28	0.22	12.00	3.42	2.73	"
Rocky Shoal Creek....	" "	2.00	0.23	0.18	10.00	2.30	1.80	"
Allen's Fork.........	County Line.........	22.52	2.56	2.04	10.00	25.60	20.40	"
Ponl Fork...........	Mangun's Mill........	10.58	1.20	0.96	9.00	10.80	8.64	"
HARALSON COUNTY.								
Tallapoosa...........	Waldrop's...........	49.80	5.60	4.48	10.00	56.00	44.80	Low spring.
" "	McBride's Bridge......	586.80	66.56	53.24	10.00	66.56	53.24	Above " "
" "	Lathrom's Crossing....	105.00	11.93	9.53	10.00	115.2	95.30	" " " "

A PARTIAL LIST OF THE WATER-POWERS IN GEORGIA, ETC.—(continued.)

Name of Stream.	Point of Section	Cubic feet per second.	Theoretical horse-power of one-foot head.	Available horse-power of one-foot head.	Approximate head or an assumed head of 10 feet.	Theoretical power of stream with this head running 24 hours.	Available power of stream with this head working 24 hours of each day.	Cond'n of Str.	By whom surveyed.	Remarks.
Haralson Co.—*Cont.*										
Little River.....	Mouth.........	19.48	2.22	1.77	10.00	22.20	17.70	Above l. sp.		
Beach Creek.........	Rock House.........	30.50	3.31	2.64	10.00	33.10	26.40	Low water.		A 30-foot dam would flood 70 acres or more.
Renfroe's Creek.	Near mouth, near Draketown	31.40	3.56	2.85	10.00	35.60	28.50	Above " "		
Harris County.										
Mulberry Creek.	Emery's Mill......	...	...	...	60.00	...	...			Too full for measurement, has about 150 feet in spring months. Falls 60 ft. in ¼ mile.
Mountain Creek......	River Road......	63.00	7.18	5.74	20.00	143.60	114.88	Low spring or more.		
Heard County.										
Potato Creek.........	County Line......	22.00	2.52	2.01	10.00	25.20	20.10	Low spring.	Locke.	Sand Beds.
New River...	¼ mile mouth...	136.08	15.68	12.54	10.00	156.80	125.40	" "	"	"
Chattahoochee........	Lot 344 and 3d...	3000.00	340.80	272.64	10.00	3408.00	2726.40	Low water estimat'd.	"	Shoals about 1 mile long.
Central Hatchee.......	Near mouth.......	100.00	11.34	9.08	10.00	113.40	90.50	Low spring.		
Jackson County.										
Curry's Creek.........	Near Jefferson...	8.00	0.91	0.72	18.00	16.42	13.13		Barrow.	

Oconee River	Hurricane Shoals	91.39	10.42	8.33	26.00	270.87	216.69			Head is all shoal.
JASPER COUNTY.										
Ocmulgee River	Lloyd's Shoals	2166.00	246.00	196.80	39.62	2840.00	7872.00			
"	Roach's Shoals	2166.00	246.00	196.80	7.50	1845.00	1476.00			
"	Barnes' Shoals	1416.00	160.80	128.64	11.64	1851.50	1481.20			
"	Seven Islands Shoals	2917.00	331.37	265.09	19 51	6620.00	5296.00			
JEFFERSON COUNTY.										
Limestone Creek	Turver's Mill	20.00	2.98	1.82	7.00	15.96	12.76		Barrow.	
Williamson Swamp	No. 11 C. R.R.	100.00	11.36	9.12	10.00	113.60	91.		"	
"	Hend's Mill	12.07	1.37	1.09	15.00	20.64	16.51			
JOHNSON COUNTY.										
Deep Creek	Parron's Mill	18.00	2.05	1.64	10.00	20.50	16.40	Above low water.	Locke.	
Buckeye Creek	7 miles from mouth	30.00	3.42	2.73	10.00	34.20	27.30	"	"	
Prong of Oboejee	Winterville Road	5.00	0.57	0.45	10.00	5.70	4.50	"	"	
JONES COUNTY.										
Ocmulgee River	Harris' Shoals	2917.00	331.37	265.09	2.30	761.30	609.00	Low water.	Probell.	
"	Johnston's Shoals	2917.00	331.37	265.09	5.10	1688.10	1320.50	"	"	
"	Holman's Shoals	2917.00	331.37	265.09	1.30	441.60	333.28	"	"	
"	Glover's Mill Shoals	2917.00	331.37	265.09	17.90	5958.00	4765.40	"	"	
LINCOLN COUNTY.										
Little River	Dill's Mill	100.00	11.36	9.12	9.00	102.6	82.06		Barrow.	
LUMPKIN COUNTY.										
Jones Creek	234, 5 and 1	5.00	0.57	0.45	50.00	28.50	22.80			Fall exclusive of dam.
Kimble Will	10 miles Dahlonega	50.00	5.70	4.56	12.00	68.40	54.72			"
owah River	5 "	200.00	29.80	18.24	10.00	223.00	182.40			"

A PARTIAL LIST OF THE WATER-POWERS IN GEORGIA, ETC.—(continued.)

Name of Stream.	Point of Section.	Cubic feet per second.	Theoretical horse-power of one foot head.	Available horse-power of one-foot head.	Approximate head or an assumed head of 10 feet.	Theoretical power of stream with this head running 24 hours.	Available power of stream with this head working 24 hours of each day.	Condition of Stream.	By whom surveyed.	Remarks.
LUMPKIN CO.—*Cont.*										
Cane Creek...	Near Dahlonega	40.00	56	3.64	10.00	45.00	36.40		Barrow.	Very large power, uses only 90 H P.
Yahoola River	Mining Co...	...	...	...	...	...	...		"	
McDUFFIE COUNTY.										
Sweet Water Creek	Cotton Card Factory	...	...	...	21.00	36.00	...		"	Estimated from wheel.
Little River	Belknap Smith	47.00	5.35	4.28	9.00	42.86	34.28		"	
MILLER COUNTY.										
Spring Creek	Colquitt	66.56	7.52	6.01	10.00	75.20	60.10	Low water.	Locke.	Banks very flat
MILTON COUNTY.										
Four Killer	Cr. Camp's Mill	28.60	2.63	2.12	20.00	53.00	42.40	Flush.	"	At low water about 10.0 cubic feet.
Big or Vickery's Creek	Above Lebanon Mills	114.39	12.95	10.32	10.00	129.50	103.20	Low spring.	"	
Little River	Graham's Mill	119.00	13.51	10.80	10.00	135.10	108.00	" or more.	"	
MONROE COUNTY.										
Bushy Creek	4 miles Danielsville	5.00	0.57	0.45	10.00	5.70	4.50	Low spring or more.	"	
Ocmulgee River	Taylor's Shoal	2917.00	331.37	265.09	5.70	1856.70	1509.30	Low water.	Probell.	Fall exclusive of dam.

Ocmulgee River	Falling Creek Shoal	2917.00	331.37	265.09	1.71	562.70	450.16	Low water.	Barrow.
" "	Dane's Shoal	2917.00	331.37	265.09	3.6	1191.00	913.98	" "	"
" "	Capp's Shoal	2917.00	331.37	265.09	5.60	1553.60	1482.88	" "	"
" "	Pitman's Shoal	2917.00	331.37	265.09	3.50	1158.50	926.80	" "	"
MURRAY COUNTY.									
Polecat Creek	214, 8, and 3	5.2	0.59	0.47	10.00	5.90	4.70		
Sugar "	208	15.3	1.74	1.39	10.00	17.40	13.10		
Mill "	299, 26, and 2	20.0	2.28	1.80	10.00	22.80	18.40	"	
Holly "	204, 26, and 2	20.0	2.28	1.80	10.00	22.80	18.10	"	
MUSCOGEE COUNTY.									
Bull Creek	Road to Woolfolk's	25.00	2.84	2.27	10.00	28.40	22.70	Above low water.	Very sandy.
Chattahoochee	Columbus	3000.00	340.80	272.64	106.0	36040.00	28832.00		Fall given by Capt. Bass. Cubic feet estimated.
NEWTON COUNTY.									
Yellow River	Georgia R.R. Bridge	666.	75.60	60.5	4.33	325.00	260.00	Frobell.	Fall of shoal exclusive of dam.
" "	Cedar Shoals	716.	81.30	65.00	63.66	5020.00	4056.00	"	"
" "	Indian Fishery Shoals	716.	81.30	65.00	12.27	996.00	796.80	"	"
" "	Allen's Shoals	716.	81.30	65.00	1.83	126.00	100.80	"	"
" "	Lee's Shoal	716.	81.30	65.00	3.97	324.00	259.20		"
" "	Dried Indian Shoal	716.	81.30	65.00	7.24	573.00	458.40	"	"
OGLETHORPE COUNTY.									
Long Creek	4 miles South Lexington	7.20	0.83	0.66	10.00	8.30	6.60	Barrow.	
PAULDING COUNTY.									
Tribut'y Pumpkinvine.	Stearn's Mill	6.00	0.68	0.54	12.0	8.16	6.52	Low spring. Locke.	
Lit. "	16 miles Marietta	10.00	1.14	0.91	20.0	22.8	18.24	" "	"
Raccoon Creek	Chappel's Store	22.0	2.51	2.00	12.0	30.00	24.0	" "	Or flush.

A PARTIAL LIST OF THE WATER-POWERS IN GEORGIA, ETC.—(continued.)

Name of Stream.	Point of Section.	Cubic feet per second.	Theoretical horse-power of one-foot head.	Available horse-power of one-foot head.	Approximate head or an assumed head of 10 feet.	Theoretical power of stream running with this head 24 hours.	Available power of stream with this head working 24 hours of each day.	Condition of Stream.	By whom surveyed.	Remarks.
PAULDING CO.—*Cont.*										
Peggymore...	Lee's, near mouth	11.18	1.26	1.01	10.00	12.60	10.10		Locke.	
Sweet Water	Seal's Bridge	12.00	1.36	1.08	10.00	13.60	10.80	Low water.	"	
PICKENS COUNTY.										
Big Scared Corn	Fairmount Road	11.00	1.25	1.00	10.00	12.50	10.00		Barrow.	
Little " "	and Jasper Road	4.50	0.51	0.40	10.00	5.10	4.00		"	
Talking Rock Creek	Federal Road	13.33	1.52	1.21	10.00	15.20	12.10		"	
Love's Creek	Federal Road	7.00	0.79	0.63	18.00	14.36	11.48		"	
Long Swamp	Below Forks	40.00	4.56	3.64	10.00	45.60	36.40		"	
Tribu'y of Long Swamp	Federal Road	6.00	0.68	0.54	10.00	6.80	5.40		"	
Stegall's Mill Creek	Stegall's Mill	10.00	1.14	0.91	10.00	11.40	9.10		"	
Long Swamp	Marble Quarry	23.00	2.62	2.09	10.00	26.20	20.96		"	
Fork Swamp	Jasper Road	8.11	0.92	0.73	12.00	11.08	8.86		"	
POLK COUNTY.										
Euharlee	Rockmart	25.00	2.85	2.28	10.0	28.50	22.80	Minimum low water.		
"	2 miles North Rockmart	19.00	2.15	1.72	10.00	21.50	17.30	Low spring.		
"	Hightower's Mill	5.40	.612	0.49	90.00	54.9	44.10	" "	"	

Big Spring	Rome and Van Wert Road, 2 miles Van Wert	5.00	.57	.45	10.00	5.70	4.50	Low spring.	Barrow.	
Little Cedar	Young's Mill	17.70	2.00	1.60	10.00	20.00	16.00	" "	"	
Big Spring	Cedar Town	9.60	1.08	.85	10.00	10.80	8.60	" "	"	
Gut Creek	At mouth	27.30	3.06	2.45	10.00	30.00	24.50	or more.	"	
QUITMAN COUNTY.										
Hoclarnee	Near mouth	6.00	0.68	0.54	10.00	6.80	5.44	Low water.	Locke.	
Tobehannee	1 mile S. E. Georgetown	10.00	1.14	0.91	10.00	11.40	9.12	" "	"	
RABUN COUNTY.										
Head of Stekoa	Near Clayton	3.75	0.43	0.34	30.00	12.90	10.32		Barrow.	
Creek	Mouth	30.00	3.42	2.73	12.00	41.04	32.83		"	
Wildcat Creek	"	50.0	5.70	4.56	10.00	57.00	45.60		"	
Tiger Creek	"	40.60	4.63	3.70	15.00	69.45	55.56			
RANDOLPH COUNTY.										
Roaring Branch	5 miles Fort Gaines	4.00	0.45	0.36	30.00	13.50	10.60	Low water.	Locke.	
Wakefortsee Creek	Near Chemochechobee	5.00	0.57	0.45	10.00	5.70	4.50	" "	"	Very high heads at times
RICHMOND COUNTY.										
Augusta Canal	Augusta					12000 0	...		B. Holly, Canal Engin'r.	
Little Split Creek	At mouth	12.00	1.36	1.08	8.00	10.94	8.75		Barrow.	
SCRIVEN COUNTY.										
Beaver Dam Creek	Jacksonborough	87.35	9.95	7.96	7.00	69.70	53.76		"	
Briar Creek	Mill Haven	565.50	64.46	51.56	10.00	644.60	515.68		"	
Rocky Creek	Wade's Mill	12.00	1.37	1.09	5.00	6.84	5.47		"	

A PARTIAL LIST OF THE WATER-POWERS IN GEORGIA, ETC.—(*continued.*)

Name of Stream.	Point of Section.	Cubic feet per second.	Theoretical horse-power of one-foot head.	Available horse-power of one-foot head.	Approximate head or an assumed head of 10 feet.	Theoretical power of stream with this head running 24 hours.	Available power of stream running this head 24 hours of each day.	Condition of Stream.	By whom surveyed.	Remarks.
STEWART COUNTY.										
Wimberly's Branch...	Grimes & Freeman's Mill....	8.80	1.00	0.80	12.00	12.00	9.60	Low water.	Locke.	Estimated.
Hodchodkee..........	Scott's Mill............	12.00	1.35	1.08	10.00	13.50	10.80	" "	"	
TWIGGS COUNTY.										
Big Sandy..........	Myrick's Mill...........	....	....	....	8.00	....	....			Too full to measure.
TROUP COUNTY.										
Shoal Creek..........	Troup Factory............	61.00	9.28	7.38	18.00	166.14	132.91	Low water or more.	"	Estimated by wheels.
Muddy Creek..........	5½ miles LaGrange........	7.00	0.79	0.63	10.00	7.90	6.30	"	"	
Blue John..........	2½ " "	3.00	0.34	0.27	10.10	3.40	2.70	"	"	
Panther Creek......	3½ " "	25.00	2.84	2.27	10.00	28.40	22.70	"	"	
Flat Creek........	3½ " Gorham's Mills..	20.00	2.28	1.82	12.00	27.36	21.88	"	"	
Beach Creek..........	5 " LaGrange........	35.00	4.00	3.20	15.00	60.00	48.00	"	"	
Yellow Jacket..........	8½ " "	87.36	9.92	8.03	10.00	99.20	80.30	"	"	
WALKER COUNTY.										
Fork of Dry Creek......	½ mile mouth........	6.5	0.74	0.59	10.00	7.40	5.90		Barrow.	

LIST OF WOODY PLANTS OF GEORGIA.

NO.	NO. OF FAM'Y.	FAMILY.	BOTANICAL NAME. GENUS. SPECIES.	COMMON NAME.	COUNTY.
1	2	Magnoliaceæ.	Illicium Floridanum.	Anise Tree.	
2	2	"	Magnolia grandiflora.	Magnolia.	
3	2	"	" glauca.	Sweet Bay.	
4		"	" umbrella.		
5		"	" acuminata.		
6		"	" cordata.		
7		"	" Fraseri.		
8		"	" Macrophylla.		
9		"	Liriodendron tulipefera.	White Poplar.	Murray.
10	3	Anonaceæ.	Asinima triloba.	Papaw.	Murray.
11	3	"	" grandiflora.		
12	24	Tiliaceæ.	Tilia Americana.	American Lime.	
13		"	" pubescens.		
14	25	Camilliaceæ.	Gordonia lasianthus.	Loblolly Bay.	
15		"	" pubescens.		
16		"	Stuartia Virginica.		
17		"	" pentagyna.	[Toothache Tree.	
18	34	Rutaceæ.	Xanthoxylum Carolinianum.	Prickly Ash or	
19		"	Ptilea trifoliata.	Hop Tree.	
20	37	Anacardiaceæ.	Rhus typhina.		
21	37	"	" glabra.		
22	37	"	" copallina.	Sumach.	Murray.
23	37	"	" pumilla.		
24	37	"	" { venenata	Poison Elder.	
25	37	"	" { toxicodendron. }	Poison Oak.	
26	37	"	" Aromatica.		
27	35	Vilaceæ.	Vitis labrusca.	Fox Grape.	
28	38	"	" æstivatis.	Summer Grape.	
29	38	"	" cordifolia.	Frost Grape.	Murray.
30	38	"	" vulpina.	Muscadine or Bullace.	
31	38	"	Ampelopsis quinquefolia.	Virginia Creeper.	
32	39	Rhamnaceæ.	Birchimia volubilis.	Supple Jack.	
33	39	"	Rhamnus lanceolatus.	Buckthorn.	
34		"	Trangula Caroliniana.	Carolina Buckthorn.	
35	40	Celastraceæ.	Euonymus Americanus.	Strawberry Bush.	
36	40	"	" atropurpuria.		
37	41	Staphylaceæ.	Staphyla trifolia.	Bladder-nut.	
38	42	Sapindaceæ.	Sapindus marginatus.	Soapberry.	
39	42	"	Æsculus glabra.	Horse-chestnut.	
40	42	"	" pavia.	Buckeye.	Whitefield
41	42	Sapindaceæ.	Sapindus flora.		
42	42	"	Asculus pariflora.		
43	43	Aceraceæ.	Acer Pennsylvanicum.	Striped Maple.	
44		"	" spicatum.	Mountain Maple.	
45		"	" saccharinum.	Sugar Maple.	
46		"	" dasycarpum.	Silver Maple.	Murray.
47		"	" aubrum.	Red or Swamp Maple.	
48		"	Negund accroides.	Ash-leaved Maple.	
49	47	Leguminoceæ.	Amorpha herbacia.		
50	47	"	" canescens.		
51	47	"	Robinia pseudacaia.	Locust.	
52	47	"	" viscosa.		
53	47	"	" hispida.		

LIST OF WOODY PLANTS OF GEORGIA.—(*Continued.*)

NO.	NO. OF FAM'Y.	FAMILY.	BOTANICAL NAME. GENUS. SPECIES.	COMMON NAME.	COUNTY.
54	47	Leguminoceæ.	Whistaria frutescens.		
55	47	"	Erythrina herbacia.		
56	47	"	Cladrustis tinctoria.	Yellow Wood.	
57	47	"	Circis Canadensis.	Red Bud.	Murray.
58	47	"	Gleditschia triacanthos.		
59	47	"	" monosperma.		
60	48	Rosaceæ.	Chrysobalanus oblorigifolius.		
61	48	"	Prunus Americana.		
62	48	"	" umbellata.		
63	48	"	" serotina.	Wild Cherry.	Murray.
64	48	"	" Virginiana.		
65	48	"	" Carolinacana.	Mock Orange.	
66	48	"	Cratægus spathulata.	Hawthorn.	
67	48	"	" æstivalis.	Summer or Red Haw.	
68	48	"			
69	48	"	7 other species.		
70	48	"	Pyrus coronaria.		
71	48	"	" angustifolia.		
72	48	"	" anarbulifolia.		
73		"	" Americana.		
74		"	Amelanchier Canadensis.		
75	49	Calycanthaceæ.	Calycanthus Floridus.		
76	49	"	" lævigatus.		
77	49	"	" glaucus.		
78	52	Lythraceæ.	Neseæ verticillata.		
79	57	Grossulaceæ.	Ribes.		
80	64	Saxifrageæ.	Hydrangea arborescens.		
81	64	"	" radiata.		
82	64	"	" quercifolia.		
83	64	"	Decumaria Barbara.		
84			Philadelphus grandiflorus.	Syringa.	
85	65	Hamamalaceæ.	Hamamelis Virginica.	Witch Hazel.	Murray.
86	65	"	Fothergilla alnifolia.		
87	65	"	Liquidambar styracifiua.	Sweet Gum.	Murray.
88	68	Cornaceæ.	Cornus alterniflora.		
89	68	"	" stricta.		
90	68	"	" paniculata.		
91	68	"	" sericea.		
92	68	"	" asperifolia.		
93	68	"	" Florida.	Dogwood.	Whitefield.
94	68	"	Nyssa multiflora.	Sour Gum.	Murray.
95	68	"	" agnatica.		
96	68	"	" uniflora.		
97	68	"	" capitata.	Ogecchee Lime.	
98	69	Capsifoliaceæ.	Symphoricarpus vulgaræ.	Snowberry.	
99	69	"	Sambucus Canadensis.	Elder.	
100	69	"	Vibernum prunifolium.		
101		"	" lentago.		
102		"	" obovatum.		
103		"	" accrifolium.		
104	69	"	" nudum.		
105	69	"	" dentatum.		
106	69	"	" scabrellum.		
107	70	Rubiaceæ.	Cephalanthus occidentalis.	Button-bush.	

LIST OF WOODY PLANTS OF GEORGIA.—(*Continued.*)

NO.	NO. OF FAM'Y.	FAMILY.	BOTANICAL NAME. GENUS. SPECIES.	COMMON NAME.	COUNTY.
108	70	Rubiaceæ.	Pinckneya pubens.	Georgia Bark.	
109	70	"	Gelsemium sempervirens.	Yellow Jessamine.	
110	76	Ericaceæ.	Gaylussaciæ frondosa.	Huckleberry.	
111	76	"	" dumosa.		
112	76	"	" resinosa.		
113	76	"	Vaccinium crassifolium.	Huckleb'ry, Bluc-	
114	76	"	" stamineum.	[berry.	
115	76	"	" arboreum.		
116	76	"	" nitidum.		
117	76	"	" myrsinites.		
118	76	"	" tenellum.		
119	76	"	" Elliottii.		
120	76	"	" corymbosum.		
121	76	"	Leucothoö axillaris.		
122	76	"	" catesbæi.		
123	76	"	" acuminata.		
124	76	"	" racemosa.		
125	76	"	Andromeda ferruginea.		
126	76	"	Oxydendrum arboreum.	Sour Wood or Sor-	
127	76	"	Clethra.	[rel Tree. Murray.	
128	76	"	Kalmia latifolia.	Calico Bush.	
129	76	"	" angustifolia.	Sheep Laurel.	Murray.
130	76	"	Rhododendron arborescens.	Roseboy Honey-	
131	76	"	" maximum.	[suckle.	
132	78	Aquifoliaceæ.	Ilex opaca.	Holly.	Murray.
133	78	"	" dahoon.		
134	78	"	" cassine.		
135	78	"	" ambigua.		
136	79	Styraceaceæ.	Styrax pulverulentum.	Storax.	
137	79	"	" grandifolium.		
138	79	"	" Amer. canum.		
139	79	"	Halesia diptera.	Snowdrop Tree.	
140	79	"	" tetraptera.		
141	79	"	Symplocos tinctoria.		
142	80	Cyrillaceæ.	Cyrilla racemiflora.		
143	80	"	Cliftonia ligustrina.	Titi.	
144	80	"	Elliottia racemosa.		
145	81	Ebenaceæ.	Dyospyros Virginiana.	Persimmon.	Murray.
146	82	Sapotaceæ.	Bumelia canuginosa.		
147	89	Bignoniaceæ.	Bignonia capreolata.	Crossvine.	Murray.
148	99	"	Tecomia radicans.	Trumpet Flower.	
149	89	"	Catalpa bignonioides.		
150	93	Verbenaceæ.	Lantana camara.		
151	93	"	Calicarpa Americana.	French Mulberry	Murray.
152	104	Oleaceæ.	Olea Americana.	Olive.	
153	104	"	Chionanthus Virginica.	Fringe Tree.	
154	104	"	Fraxinus Americana.	White Ash.	Murray
155	104	"	" pubescens.	Red Ash.	
156	104	"	" viridis.	Green Ash.	
157	104	"	" platycarpa.	Water Ash.	
158	104	"	Foresiera ligustrina.		
159	111	Lauraceæ.	Persea Carolinensis.	Red Bay.	
160	111	"	Sassafras officinale.	Sassafras.	
161	111	"	Benzoin odoriferum.	Spice Bush	

LIST OF WOODY PLANTS OF GEORGIA.—(*Continued.*)

NO.	NO. OF FAM'Y.	FAMILY.	BOTANICAL NAME. GENUS. SPECIES.	COMMON NAME.	COUNTY.
162	111	Lauraceæ.	Tetranthera geniculata.		
163	112	Thymeleaceæ.	Dirca palustris.	Leatherwood.	Murray.
164	124	Moraceæ.	Morus rubra.	Mulberry.	Murray.
165	125	Ulmaceæ.	Ulmus fulva.	Slippery Elm.	Murray.
166	125	"	" Americana.	Elm.	"
167	125	"	" alata.	Wahoo.	"
168	125	"	Planera aquatica.	Planer Tree.	
169	125	"	Celtis occidentalis.	Nettle Tree.	
170	126	Platanaceæ.	Platanus occidentalis.	Sycamore.	Whitefield.
171	127	Juglandaceæ.	Carya alba.	Shell-bark Hick-	"
172		"	" tomentosa.	Hickory. [ory.	"
173		"	" glabra.	Pig-nut.	"
174		"	" amara.	Butternut.	
175		"	Juglans nigra.	Black Walnut.	
176		"	" cinerea.	Butternut.	
177	128	Cupuliferæ.	Quercus phellos.	Willow Oak.	
178	128	"	" cinerea.	High-ground Oak.	
179	128	"	" virens.	Live Oak.	
180	128	"	" aquatica.	Water Oak.	
181	128	"	" nigra.	Black Jack.	
182	128	"	" catesbæi.	Turkey Oak.	
183	128	"	" tinctoria.	Black Oak.	Whitefield.
184	128	"	" coccinea.	Scarlet Oak.	
185		"	" rubra.	Red Oak.	Whitefield.
186	128	"	" Georgiana.	Stone Mt. Oak.	
187	128	"	" falcata.	Spanish Oak.	•
188	128	"	" ilicifolia.	Bear Oak.	
189	128	"	" obtusiloba.	Post Oak.	Whitefield.
190	128	"	" alba.	White Oak.	"
191	128	"	" lyrata.	Overcup Oak.	
192	128	"	" prinus.	Swamp Chestnut.	
193	128	"	" prinus.	Chestnut Oak.	
194	128	"	" prinoides.	Chinquapin Oak.	
195		"	" Castanea Americana.	Chestnut.	Whitefield.
196		"	Castanea pumila.	Chinquapin.	
197		"	Fagus ferruginea.	Beech.	Murray.
198		"	Coryllus Americana.	Hazel-nut.	
199		"	" rostrata.	Beaked Hazel-nut.	
200		"	Carpinus Americana.	Hornbeam.	Whitefield.
201		"	Ostrya Virginica.	Hop Hornbeam.	
202	129	Myricaceæ.	Myrica cerifera.	Wax Myrtle.	
203	129		" inodora.		
204	130	Betulaceæ,	Betula nigra.	Black Birch.	
205	130	"	" lenta.	Cherry Birch.	
206	130	"	Alnus serrulata.	Alder.	
207	131	Salicaceæ.	Salix tristis.	Sage Willow.	
208	131	"	" humilis.		
209	131	"	" nigra.		Whitefield.
210		"	Populus angulata.		
211		"	" grandidentata.		
212		"	" heterophylla.	Cotton-wood.	
213	132	Coniferæ.	Pinus pungens.		
214	132	"	" inops.	Scrub Pine.	
215	132	"	" glabra.	Spruce Pine.	**Murray.**

LIST OF WOODY PLANTS OF GEORGIA.—*(Continued).*

NO.	NO. OF FAM'Y.	FAMILY.	BOTANICAL NAME. GENUS. SPECIES.	COMMON NAME.	COUNTY.
216	132	Coniferœ.	Pinus mitus,	Short-leaued Pine,	Murray.
217	132	"	" rigida,	Pitch Pine,	
218	132	"	" serotina,	Pond Pine,	
219	132	"	" tœda,	Loblolly Pine,	Whitefield..
220	132	"	" australis,	Long-leaved Pine,	
221	132	"	" strobus,	White Pine,	Murray.
222	132	"	Abies Canadensis,	Hemlock Spruce,	
223	132	"	Juniperus Virginiana,	Red Cedar,	
224	132	"	Cupressus thyoides,	White Cedar,	
225	132	"	Taxodium distichum,	Cypress,	
226		"	Torreya taxifolia,		
227	134	Palmaccœ,	Sabal palmetto,		
228	131	"	" serrulata,		
229	134	"	Chamærops hystrix		
230	134	"	Prunus spinosa,	Bullace Plum, Sloe.	

No. 131.—Number and nationality of immigrants arrived in the United States during the nine years ended June 30, from 1871 to 1879, inclusive.

COUNTRIES.	1871.	1872.	1873.	1874.	1875.	1876.	1877.	1878.	1879.	Total.
England	56,630	69,764	74,801	50,905	40,130	24,373	19,161	18,405	24,183	378,252
Ireland	57,439	68,782	77,344	53,707	37,967	19,575	14,569	15,932	20,013	365,268
Scotland	11,984	13,916	13,841	10,429	7,310	4,552	4,135	3,502	6,226	74,924
Wales	809	1,214	840	665	449	324	281	243	543	5,458
Jersey Island		4	13	5		1				23
Guernsey Island				1		1				3
Channel Islands					8					8
Isle of Man		11	4	16	6	10	4			51
Great Britain, not specified	16,042								3	16,045
Total, British Isles	142,894	153,641	166,843	115,728	85,881	48,866	38,150	32,082	49,667	840,032
Germany	82,554	141,109	149,671	87,291	47,769	31,937	29,298	29,313	34,602	633,544
Austria	4,884	4,182	5,766	7,888	6,882	5,646	5,023	4,504	5,331	50,106
Hungary	3	228	1,347	962	776	630	373	646	632	5,597
Sweden	10,699	13,464	14,303	5,712	5,573	5,603	4,991	5,390	11,001	76,736
Norway	9,418	11,421	16,247	10,384	6,093	5,173	4,588	4,769	7,345	75,428
Denmark	2,015	3,690	4,931	3,082	2,656	1,647	1,695	2,106	3,474	25,195
Netherlands	993	1,909	3,811	2,444	1,237	866	591	608	753	13,201
Belgium	774	738	1,176	817	615	515	488	354	512	5,989
Switzerland	2,269	3,650	3,107	3,093	1,814	1,549	1,686	1,808	3,161	22,137
France	3,137	9,317	14,798	9,643	8,321	8,002	5,856	4,159	4,655	67,888
Italy	2,805	4,144	8,715	7,596	3,570	2,910	3,143	4,131	5,759	42,773
Sicily	11	44	41	62	61	104	52	212	32	619
Sardinia		2		8		2		1		13
Corsica	1		1	1	1					4

										Total
Malta		8	4	5	6	7	6	2	3	40
Greece	11	12	23	30	25	19	24	16	21	187
Spain	558	595	541	485	601	518	665	457	457	4,877
Portugal	290	416	24	60	763	471	1,291	660	392	4,367
Gibraltar	1	7	6	6	3	10	9	1	1	43
Russia in Europe	673	994	1,560	3,960	7,982	4,764	6,579	3,037	4,434	33,983
Poland	635	1,647	3,338	1,795	984	925	633	547	489	10,793
Finland		24	74	113	15	10	20	11	19	286
Lapland						1				1
Heligoland										6
Turkey in Europe	22	20	53	62	27	38	32	29	29	312
Europe, not specified	1									1
Total Europe, not British Isles	121,664	197,824	229,537	145,504	95,774	71,237	66,942	62,750	83,103	1,074,125
Total Europe	264,548	351,265	396,380	261,232	181,635	120,103	105,092	100,832	133,070	1,914,157

SUMMARY.

										Total
From Europe	264,518	351,265	396,380	261,232	181,635	120,103	105,092	100,832	133,070	1,914,157
From Asia	7,236	7,825	20,326	13,857	16,498	22,943	10,640	9,014	9,660	117,999
From Africa	43	62	39	22	54	54	22	14	17	327
From America	48,840	42,205	40,337	35,339	26,642	26,686	24,065	27,204	33,025	302,343
From Pacific Islands	25	2,416	1,413	1,170	1,269	1,312	914	606	816	9,941
From all other, not specified	658	1,033	1,308	1,719	1,400	888	1,124	799	1,238	10,167
Aggregate	321,350	404,806	459,803	313,339	227,498	169,986	141,857	139,469	177,826	2,354,934

No. 117.—Quantities of Raw Cotton produced, imported, exported and retained for consumption in the United States, from 1856 to 1879, inclusive.

Year ended June 30	ANNUAL CROP. Production. a (Bales)	ANNUAL CROP. Average net weight of bale. (Pounds)	ANNUAL CROP. Production in pounds, gross weight. a	Imports. (Pounds)	Total Production and Imports. (Pounds)	Exports, Domestic and foreign. (Pounds)	Retained for home consumption. (Pounds)	Percentage of Production and Imports retained for home consumption.	Percentage of Production and imports exported.
1856	3,645,345	420	1,622,907,594	1,096,841	1,624,044,435	1,361,431,701	272,572,734	16.78	83.22
1857	3,053,519	444	1,438,520,102	802,233	1,439,322,335	1,048,282,475	391,039,860	27.17	72.68
1858	3,238,962	442	1,517,518,476	690,800	1,518,109,276	1,118,624,012	399,485,264	26.31	73.69
1859	3,994,481	447	1,892,664,987	743,500	1,893,408,487	1,386,738,676	506,669,811	26.76	73.24
1860	4,823,770	445	2,275,372,309	2,005,529	2,277,377,838	1,767,830,609	509,547,229	22.37	77.63
1861	3,826,088	477	1,934,645,608	881,371	1,935,426,974	307,634,242	......	......	......
1862	b......	......	b......	29,640,853	......	5,198,230	......	......	......
1863	b......	......	b......	33,877,365	......	12,904,119	......	......	......
1864	b......	......	b......	26,475,957	......	13,420,146	......	......	......
1865	b......	......	b......	36,033,426	......	12,042,216	......	......	......
1866	2,228,987	441	1,041,962,263	6,282,341	1,048,244,604	651,921,489	396,323,115	37.18	62.19
1867	2,069,271	444	969,175,303	926,021	970,101,324	662,733,679	307,367,645	31.68	63.32
1868	2,498,895	443	1,173,431,114	514,992	1,173,946,106	785,416,226	388,530,880	33.10	66.90
1869	2,439,039	437	1,129,811,645	1,522,068	1,131,333,713	644,957,327	486,376,386	42.99	57.01
1870	3,154,946	434	1,461,401,357	1,598,133	1,463,099,490	938,785,304	494,314,186	34.02	65.98
1871	4,362,317	438	2,020,693,736	1,196,840	2,021,890,576	1,463,704,607	563,186,069	27.61	72.39
1872	2,974,351	439	1,384,084,494	2,894,183	1,536,978,677	933,825,710	463,152,967	32.67	67.33
1873	3,930,508	440	1,833,188,981	4,425,524	1,837,614,455	1,200,398,178	637,216,277	34.68	65.32
1874	4,170,388	439	1,940,648,352	3,625,830	1,944,274,182	1,358,979,913	585,294,269	30.10	69.90
1875	3,882,991	439	1,783,644,032	2,149,332	1,785,793,364	1,260,851,944	524,941,420	29.40	70.60
1876	4,669,288	436	2,157,968,142	2,451,419	2,160,409,561	1,491,629,831	668,779,730	30.96	69.04
1877	4,485,423	438	2,082,492,190	2,656,567	2,085,148,757	1,445,647,079	639,501,678	30.67	60.33
1878	4,811,265	450	2,294,973,405	3,032,013	2,298,005,448	1,608,469,052	689,536,366	30.	70.
1879	5,073,531	443	2,382,428,687	2,993,677	2,385,422,364	1,628,875,979	756,646,385	31.72	68.28

a In the columns of "Production," the amount placed opposite the fiscal year is the production of the preceding calendar year, since the exports and consumption of cotton during the fiscal year are mainly of the production of the preceding calendar year. b No record during the war period.

Chattanooga
6 03

DA[.]E
Coal
O
Trenton

WALK[.]

CARBO[.]

CHA[.]

Coo[.]

A [.] L [.]

Cedar
Bucht[.]

A D

Map of

ORGIA,

WITH

GICAL OUTLINES.

GE LITTLE, Ph.D.,
State Geologist.

S. SCHLEY, M.E., Del.

EXPLANATION.

County Seats ⊙

Railroads ⊣⊢⊢⊢⊢⊢⊢⊢⊢⊢⊢

" projected ....